Resource Management for On-Demand
Mission-Critical Internet of Things
Applications

Resource Management for On-Demand Mission-Critical Internet of Things Applications

Junaid Farooq
University of Michigan-Dearborn

Quanyan Zhu
New York University

IEEE PRESS

WILEY

Registered Office
John Wiley & Sons, Inc., 111 River Street, Hoboken, NJ 07030, USA

Editorial Office
111 River Street, Hoboken, NJ 07030, USA

For details of our global editorial offices, customer services, and more information about Wiley products visit us at www.wiley.com.

Wiley also publishes its books in a variety of electronic formats and by print-on-demand. Some content that appears in standard print versions of this book may not be available in other formats.

Library of Congress Cataloging-in-Publication Data Applied for:
ISBN: 9781119716099

Cover Design: Wiley
Cover Image: © Vasin Lee/Shutterstock

Set in 10/12pt WarnockPro by Straive, Chennai, India

10 9 8 7 6 5 4 3 2 1

To our parents

Contents

Preface

Modern infrastructure is becoming increasingly connected and autonomous, giving rise to new paradigms such as smart homes, offices, factories, and cities, etc. These utilize recent advancements in wireless communication technologies, embedded systems, and artificial intelligence-based software to create an ecosystem of networked electronic devices, referred to an Internet of things (IoT). The cyber-physical integration, enabled by the IoT, is resulting in a myriad of applications and services. However, the IoT itself is not a standalone system. Instead, it is composed of a variety of different systems and components such as endpoint devices, communication networks, cloud computing systems, and user devices, etc. Furthermore, the footprint of the IoT is massive and the constituent components are often operated and controlled by completely different entities. Therefore, effective management and operation of the resources in the IoT ecosystem requires the development of policies, decision-making frameworks, and practical mechanisms.

At each level of interactions between the systems in the IoT, there are decision problems that arise for achieving various objectives such as enhancing the efficiency, quality of service and experience, economics, security, and resilience, etc. This book takes a holistic cyber-physical view towards decision-making, when applied to large scale IoT systems and networks that may also be highly dynamic in nature. For instance, at the device level, it is important to ensure wireless connectivity and communication efficacy. If the communication infrastructure such as wireless base stations are deployed, there is a need for regulating the spectrum usage. However, in the absence of such infrastructure, the connectivity might have to be achieved using alternative techniques such as an overlay network of aerial base stations. Decisions have to be made on how to allocate the spectrum to users and how to configure and place aerial base stations to provide adequate coverage and connectivity. This book proposes a reservation-based spectrum usage mechanism that provides performance guarantees to enable real-time and mission-critical applications. In the absence of backhaul wireless connectivity, a novel approach based on flocking of unmanned aerial vehicles is proposed for providing uninterrupted wireless services to ground users and communities.

Once the connectivity of devices has been achieved, networks are used for information dissemination. The emphasis is on designing networks that achieve the desired information propagation and are reconfigurable in the case of adversarial attacks. Furthermore, there is a need to design security mechanisms against stealthy adversarial threats that may be using the same communication networks to infiltrate and sabotage network operation. This book proposes a novel modeling and analysis approach based on spatial point processes and mathematical epidemiology to analytically characterize the propagation of information and/or malware over wireless communication networks. This is then used to propose decision support in designing networks that are geared towards achieving the desired objective such as information dissemination or network security. The developed theoretical foundations have widespread applications in military and civilian scenarios.

The next frontier in the IoT ecosystem is the resource allocation and service provisioning, which appears in a variety of scenarios such as the use of cloud computing resources by smart devices or the use of city resources in an urban setting. Service requests may appear randomly over time and space with varying service requirements. It is important to efficiently allocate and price the available resources in order to provide a high quality of experience to the users and to generate high revenues for the cloud service provider. This book proposes revenue maximizing approaches to filter the requests dynamically and sets rules for allocation and pricing of available resources. The decisions are based on statistical modeling of the service requests and predictive analysis to schedule and assign resources to incoming requests.

The overarching goal of this book is to lay the theoretical foundations of decision and management science in IoT network design and operation. It leverages tools and theories from a diverse range of systems sciences such as mathematical epidemiology, spatial point processes, stochastic processes, optimal control theory, and optimization to address the challenges and problems at multiple levels across the IoT stack. In a nutshell, this book aims to close the gap between the theory of dynamic mechanism design and wireless and IoT systems. It also paves the way for the development of a comprehensive science for decision-making in the IoT networks. The proposed mechanisms in this book are envisioned to enable and support the development of smart and connected cities with features including resilient communications infrastructure, next generation emergency response, critical infrastructure security, sensing and data markets, etc. We hope that this book will provide a broad understanding of the multiple facets of resource management in the IoT ecosystem and will enable the readers to delve deeper in this interesting and rapidly evolving area of research.

Dearborn, MI, March 2021
Brooklyn, NY, March 2021

Junaid Farooq
Quanyan Zhu

Acknowledgments

We would like to acknowledge the support that we receive from our institutions: the University of Michigan-Dearborn and New York University (NYU). We also thank many of our friends and colleagues for their inputs and suggestions. Special thanks go to the members of the Laboratory of Agile and Resilient Complex Systems (LARX) at NYU, including Jeffrey Pawlick, Juntao Chen, Rui Zhang, Tao Zhang, Linan Huang, Yunhan Huang, and Guanze Peng. Their encouragement and support has provided an exciting intellectual environment for us where the major part of the work presented in this book was completed. We would also like to acknowledge support from several funding agencies, including the National Science Foundation (NSF), Army Research Office (ARO), and the Critical Infrastructure Resilience Institute (CIRI) at the University of Illinois at Urbana-Champaign for making this work possible.

Junaid Farooq and Quanyan Zhu

Acronyms

AI	Artificial Intelligence
AP	Access Point
AR	Augmented Reality
BS	Base Station
C^4ISR	Command, Control, Communications, Computers, Intelligence, Surveillance, and Reconnaissance
CDF	Cumulative Distribution Function
CI	Critical Infrastructure
CPS	Cyber-Physical Systems
CRN	Cognitive Radio Networks
CSMA	Carrier Sense Multiple Access
CSP	Cloud Service Provider
D2D	Device-to-Device
DDoS	Distributed Denial of Service
FSD	First-Order Stochastic Dominance
HVAC	Heating Ventilation and Air Conditioning
IC	Incentive Compatibility
IoBT	Internet of Battlefield Things
IoT	Internet of Things
IR	Individual Rationality
LPWA	Low-Power Wide-Area
MAC	Medium Access Control
MAP	Mobile Access Point
MC-IoT	Mission-Critical Internet of Things
MD	Mechanism Design
MSD	Mobile Smart Device
PDF	Probability Density Function
PMF	Probability Mass Function
PPP	Poisson Point Process
QoE	Quality of Experience
QoS	Quality of Service

RAN	Radio Access Network
RGG	Random Geometric Graph
RTT	Round Trip Time
SG	Stochastic Geometry
SINR	Signal-to-Noise-Plus Interference-Ratio
SIS	Susceptible-Infected-Susceptible
TF	Time-Frequency
ToI	Things of Internet
UAV	Unmanned Aerial Vehicle
VM	Virtual Machine
VMI	Virtual Machine Instance
VR	Virtual Reality
WAN	Wide Area Network
WDoS	Wireless Denial of Service
WPAN	Wireless Personal Area Network

Part I

Introduction

1

Internet of Things-Enabled Systems and Infrastructure

1.1 Cyber–Physical Realm of IoT

Network-connected electronic devices are becoming an essential part of modern infrastructure systems to automate manual processes resulting in improved efficiency and productivity. The Internet of Things (IoT) is an interconnection of different types of devices (classified as sensors and actuators) using communication networks and computing systems to achieve such automated operation. The difference in IoT from traditional computing systems is their interaction with the physical world as opposed to just the cyber world. For instance, we have electronic devices controlling the temperature in smart buildings by sensing the environment and operating the heating, ventilation, and air conditioning (HVAC) systems. The IoT is in fact a massive network of cyber–physical systems (CPSs). Therefore, the cyber and physical components are an integral part of the emerging IoT ecosystem. The cyber and physical systems are coupled together in an intricate fashion where the cyber world influences decisions in the physical world and vice versa.

Figure 1.1 shows the structure of a typical IoT system. In essence, there are several actors involved in setting up the IoT ecosystem that includes sensing/actuating devices, firmware, radio access network (RAN), cloud server, mobile apps, and end user devices. The endpoint devices are made of embedded hardware that interact with the physical environment and are driven by software processes referred to as firmware or operating system. They make use of communication infrastructure, which is composed of access points, gateways, and core IP networks to connect to cloud servers, that in turn host applications and services, which are operated by users via computing devices, such as smart phones, smart watches, and voice assistants, etc.

Resource Management for On-Demand Mission-Critical Internet of Things Applications, First Edition.
Junaid Farooq and Quanyan Zhu.
© 2021 John Wiley & Sons, Inc. Published 2021 by John Wiley & Sons, Inc.

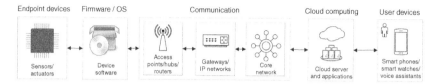

Figure 1.1 IoT technology stack describing the different actors involved in setting up the IoT ecosystem.

1.2 IoT in Mission-Critical Applications

IoT systems have a wide variety of application areas. Some of the IoT applications are highly delay-sensitive, e.g. real-time systems such as those involving artificial intelligence (AI), virtual reality (VR) and augmented reality (AR), real-time control loops, streaming analytics, etc. [121]. Such applications are referred to as *mission-critical* [158] not only due to conventional "life risk" interpretation but also pertaining to the risks of interruption of public services interruption, perturbing public order, jeopardizing enterprise operation and causing losses to businesses, etc. In mission-critical IoT (MC-IoT) applications, often a delay in communication may in fact fail the initial objective of the application. For instance, in a surveillance system where an unusual activity needs to be reported promptly to avoid any potential damage or loss of property and a report beyond a certain delay may be futile. Nevertheless, the traditional MC definition still holds and more so since IoT is also being rapidly integrated into these systems such as in public safety systems or other emergency networks [38] requiring dedicated resources at all times due to the unpredictability of unforeseen events.

1.3 Overview of the Book

The book is organized into six main parts. Part I provides a high level description of the IoT ecosystem and its main features; Par II provides an overview of the main design challenges facing the IoT systems and networks across different layers; Part III investigates and addresses the wireless connectivity challenge faced by the endpoint devices; Part IV tackles the networking layer challenges including information dissemination and message propagation over networks. Part V deals with the service provisioning and resource allocation problems in IoT for mission-critical service delivery. Finally, Part VI provides a broader view of the impact of this work along with a vision governing future research in this domain. Each part is further organized into multiple chapters. The main contributions of the book along with references to the relevant chapters are summarized in the following subsection.

1.3.1 Main Topics

This book takes a clean slate approach toward the design of IoT-enabled systems and networks. It uses a cross-layer perspective in decision-making across various avenues in the IoT ecosystem. Some of the main topics are presented in the following subsections.

1.3.1.1 Dynamic Reservation of Wireless Spectrum Resources

In Chapter 5, a dynamic mechanism for spectrum reservation considering the uncertainty in available spectrum at each time and the uncertainty in the requirement for spectrum access is developed. We make use of tools from sequential screening [26] and mechanism design literature to establish a dynamic menu of contracts which comprise of an advanced payment for spectrum reservation in the future along with a rebate policy if the spectrum is released before the time of spectrum access. This allows the network operator to discriminate the unknown application types and generate higher profits than the traditional auction mechanisms where every application is completely aware of its true utility. A two-type categorization of IoT applications is considered where they are classified as either MC or non-MC and consequently an optimal binary contract is designed by the service provider. Based on assumptions on the distribution of utility of the MC and non-MC applications, closed form results for the optimal contracts are derived and the effect of system parameters is analyzed to gain insights.

1.3.1.2 Dynamic Cross-Layer Connectivity Using Aerial Networks

In Chapter 6, a dynamic approach is used to configure robotic network nodes such as unmanned aerial vehicles (UAVs) to provide connectivity to IoT devices. Although the existing methodologies provide optimization based approaches to the UAV placement problem, this problem is dynamic in nature and hence a more holistic approach is required to obtain an efficient placement of the UAVs in real-time. In addition to effective initial deployment of UAVs, there is a need for an autonomic, self-organizing, and self-healing overlay network that can continuously adapt and reconfigure according to the constantly changing network conditions [23]. Therefore, a distributed and dynamic approach to providing resilient connectivity is essential to cope with the growing scale of the networks toward a massive IoT [44]. To this end, this book develops a feedback based distributed cognitive framework that maintains connectivity of the network and is resilient to the mobility of ground users and/or failures of the UAVs. The continuous feedback enables the framework to actively react to network changes and appropriately reconfigure the network in response to a failure event that has resulted in loss of connectivity. Simulation results demonstrate that if sufficient UAVs are available, they can be arranged into a desired configuration from arbitrary initial positions and the configuration continuously adapts according to the movement of the ground users as well as recovers connectivity under varying levels of a random UAV failure event.

1.3.1.3 Dynamic Processes Over Multiplex Spatial Networks and Reconfigurable Design

In Chapter 7, a stochastic geometry (SG) based model is used to characterize the connectivity of wireless networks in adversarial environments such as battlefields. We then use an epidemic spreading model to capture the dynamic diffusion of multiple messages within the network of devices at the equilibrium state. A novel multiplex network model for Internet of battlefield things (IoBT) networks is proposed that helps in characterizing the intra-layer and network-wide connectivity of heterogeneous battlefield devices by considering the spatial randomness in their locations. A tractable framework is developed for quantification of simultaneous information dissemination in the multiplex IoBT network based on mathematical epidemiology that considers the perceived level of threat to the network from cyber–physical attacks. Approximate closed form results relating the proportion of informed devices at equilibrium and the network parameters are provided. The resulting integrated open-loop system model is used as a basis for reconfiguring the network parameters to ensure a mission-driven information spreading profile in the network. An optimization problem is formulated that can assist military commanders in identifying the physical network parameters that are required in order to sufficiently secure the network from the perceived attacks. It can also help in reconfiguring existing networks to achieve a desired level of communication reliability. A detailed investigation of the developed integrated framework is provided for particular battlefield missions, and the effect of threat level and performance thresholds is studied. This book bridges the gap between the spatial stochastic models for wireless networks and the dynamic diffusion models in contact-based biological networks to derive new insights that aid in the planning and design of secure and reliable IoBT networks for mission critical information dissemination. The developed framework, with some modifications, is also applicable to the more general class of heterogeneous ad-hoc networks.

In Chapter 8, novel methodologies are proposed to overcome the unique challenges of modeling and analyzing the crucial interplay between malware infection, control commands propagation, and device patching in wireless IoT networks. We leverage ideas from the theories of dynamic population processes [70] and point processes to setup a mean field dynamical system that captures the evolution of malware infected devices and control command aware devices over time. In general, obtaining tractable characterizations of the equilibrium state in such population processes is theoretically involved due to the self-consistent nature of the equations involved and the complex connectivity profile of the network. However, we propose a variation of the mean field population process model based on a customized state space that allows us to analyze the formation of botnets in wireless IoT networks and helps in making decisions to control its impact.

1.3.1.4 Sequential Resource Allocation Under Spatio-Temporal Uncertainties

In Chapter 9, an adaptive and resilient dynamic resource allocation and pricing framework is developed for the context of cloud-enabled IoT systems. We present an optimal dynamic policy to filter incoming service requests by IoT applications based on the complexity of the tasks. The qualification threshold for tasks is adaptive to the number of available virtual machines (VMs), the arrival rate of requests, and their average complexity. The optimal policy can be dynamically updated in order to maintain high expected revenues of the cloud service provider (CSP). Furthermore, the proposed framework is also able to adapt according to the changing availability of the VMs due to reprovisioning of resources for other applications or due to the effect of malicious attacks.

In Chapter 10, a revenue maximizing perspective toward allocation and pricing in fog based systems designed for mission critical IoT applications is proposed. The quality-of-experience (QoE) resulting from the pairing of fog resources with computation requests is used as a basis for pricing. We develop a dynamic policy framework leveraging the literature in economics, mechanism design [51], and dynamic revenue maximization [52] to provide an implementable mechanism for dynamic allocation and pricing of sequentially arriving IoT requests that maximizes the expected revenue of the CSP. The developed optimal policy framework assists in both determining which fog node to allocate an incoming task to and the price that should be charged for it for revenue maximization. The proposed policy is statistically optimal, dynamic, i.e. adapts with time, and is implementable in real-time as opposed to other static matching schemes in the literature. The dynamically optimal solution can be computed offline and implemented in real-time for sequentially arriving computation requests.

In Chapter 11, the spatio-temporal aspect is combined with incomplete information about resource requests to devise an integrated resource provisioning framework. Ideas from the stochastic assignment of sequentially arriving tasks to workers [32] are enriched to encompass a more generic utility function that also incorporates the spatial dimension of the sequentially arriving requests. Statistical properties of utility maximizing spatial service requests are characterized using spatio-temporal Poisson processes and extreme value analysis. Analysis for a generalized utility function is done based on the distance from the source as well as the magnitude of the request. Special cases of the utility function are considered for numerical evaluations. An integrated and holistic policy framework is developed that is dynamically optimal and can act as the foundation for allocation and pricing in a wide variety of applications in the context of smart city applications. Finally, a comparison of the performance of the proposed resource provisioning framework is provided with benchmark allocation strategies.

1.3.2 Notations

The book has used a breadth of different notations for studying various theoretical models. In general, there has been an attempt to keep the notations consistent unless otherwise highlighted individually on a case-by-case basis. Some common notations used are presented as follows:

Double-struck symbols, e.g. $\mathbb{R}, \mathbb{Z}, \mathbb{N}$, etc. generally indicate sets or spaces, except $\mathbb{P}(\cdot)$, which represents the probability measure, and $\mathbb{E}[\cdot]$, which represents the expectation operator. The associated density and distribution functions are denoted by $f_X(\cdot)$ and $F_X(\cdot)$, respectively. Upper case alphabets, e.g. K, M, etc. are generally used to represent random variables, while lower case alphabets indicate the value assumed by the random variables. The operator $|\cdot|$ represents the cardinality of the set, while $\|\cdot\|$ denotes the Euclidean norm. First derivative and second derivatives are denoted by $(\cdot)'$ and $(\cdot)''$, respectively. Other notations apply to the chapters within which they are defined.

2

Resource Management in IoT-Enabled Interdependent Infrastructure

The Internet of Things (IoT) is becoming indispensable in the critical infrastructure (CI) systems such as in energy, transportation, communications, emergency services, public administration, defense, etc., due to their burgeoning scale and complexity. Furthermore, CIs may be interdependent, for instance, most CIs depend on energy systems and hence the energy systems may have a significant influence on the operation of other CIs. An illustration of the IoT-enabled infrastructure is provided in Figure 2.1, where the observations or sensing information is obtained from the actual infrastructure. This information may propagate over networks involving the cyber infrastructure. Finally, services are offered over these networks, where agents interact with each other.

2.1 System Complexity and Scale

The IoT inherently is a large-scale system due to the massive interconnectivity. This feature implies that dealing with individual devices and or systems may be difficult and prohibitive. Therefore, we need to consider characteristics and attributes that apply to a wider class of systems. It is instrumental for capturing the scale, magnitude, and complexity of systems in the modern world. In large-scale and complex systems, collective behaviors or states are more important than individual ones. In this book, the focus is on understanding the macroscopic phenomena instead of localized behaviors. While it abstracts smaller and fine grained details of the systems, it is useful in determining the systemic response and behaviors. Apart from the macroscopic view, the abstractions also assist in mathematical tractability and analysis since concepts from mean field theory become useful in modeling the interactions. This book leverages ideas from large-scale networks and mean-field analysis to study equilibrium behaviors in networks and systems that are relevant to IoT networks.

Resource Management for On-Demand Mission-Critical Internet of Things Applications, First Edition. Junaid Farooq and Quanyan Zhu. © 2021 John Wiley & Sons, Inc. Published 2021 by John Wiley & Sons, Inc.

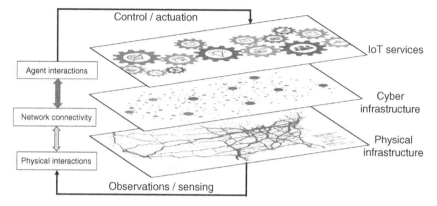

Figure 2.1 IoT-enabled infrastructure systems.

2.2 Network Geometry and Dynamics

Since the IoT relies heavily on wireless connected endpoint sensors and actuator devices, the physical distances between devices become all the more important in the evaluation of the received signal power and the interference power. Therefore, the physical network geometry plays a key role in the performance of the network. Furthermore, the locations of devices may not be fixed or even well-known. Therefore, spatial stochastic models are valuable in order to model such scenarios for analysis and decision-making. Figure 2.2 shows a randomly generated instance of node locations according to a Poisson point process representing the physical location of wireless devices in space. It also shows how

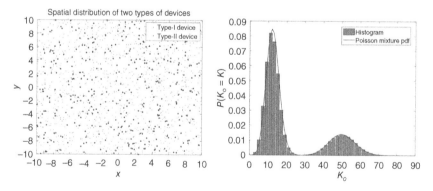

Figure 2.2 Geometry of heterogeneous wireless network and degree profile of a typical device.

having different types of devices with non-homogeneous communication range affects the degree distribution of a typical node in the network.

Moreover, the IoT systems and networks may have dynamics in several different dimensions. These dynamics may be related to the physical aspects of the networks. For instance, nodes may enter or leave the network over time. Similarly, the locations of devices may change over time because of mobility. However, the dynamics may also be in a virtual sense. For instance, the state of devices may change over time. This book focuses only on the virtual dynamics of the systems. More specifically, we consider that dynamics may be present in the availability or propagation of information in the network. An interesting feature considered in this book is the study of the dynamics together with the physical geometry of the network. This has been described in detail in Chapter 7.

2.3 On-Demand MC-IoT Services and Decision Avenues

Decision theory is an interdisciplinary approach to determine how decisions are made given unknown variables and uncertain decision environment. It brings together ideas from the field of psychology, statistics, philosophy, and mathematics to analyze the decision-making process. In an IoT ecosystem, there are several different avenues that require the use of decision theoretic ideas to efficiently design and operate the system. They can be summarized as follows:

- **Endpoint Connectivity**: Decisions have to be made on the allocation of spectrum to users if the underlying infrastructure, such as base stations, is available. However, in case of an absence of communication infrastructure, alternative arrangements need to be made such as deployment of aerial base stations using unmanned aerial vehicles (UAVs). The decision then lies in the appropriate configuration/placement of aerial base stations to provide adequate coverage and connectivity.
- **Communication/Information Dissemination**: Once the connectivity is achieved, the networks are used for information dissemination. It is important to design network parameters such that they ensure a highly reliable and network wide dissemination of information that may be critical. Furthermore, network heterogeneity and other key elements need to be considered when deploying and configuring the network.
- **Cloud/Fog Computing Infrastructure**: The next frontier is the use of cloud computing resources by smart devices. It is important to allocate and price the computational resources efficiently to provide a high quality of

experience to the users and generate high revenues for the cloud/fog service provider.

- **System Level/Policy Decisions**: At the application front, there are scenarios where resource provisioning decisions need to be made for service requests that appear randomly in space and time, for instance, in the case of smart urban environments.

2.4   Performance Metrics

In this book, the key metrics that will be used are broadly categorized into the following three areas:

- **Efficiency**: It includes objectives such as energy consumption, delay, connectivity, etc.
- **Economics**: It includes objectives such as pricing, revenue maximization, profitability, etc.
- **Security and Resilience**: It is related to the protection, reconfigurability, and self-organizing ability of systems.

These include objectives such as protection from risks of large scale distributed denial-of-service (DDoS) attacks, botnet formation, etc. The goal of this book is to embed security in the design of systems and networks. Efficiency is often the most important concern in system design. There is also a delicate trade-off between efficiency and security or reliability. It is important that systems are made secure from intentional and unintentional attacks, while maintaining high operational performance. However, no matter how secure systems are built, they may still be vulnerable to failures and attacks. The next goal of the designers is to ensure that systems fail gracefully in the event that attacks ultimately occur. Therefore, it is important to create mechanisms that are either reconfigurable or self-organizing to recover the system state from the impact of the attacks. This property of the system is referred to as resilience. Finally, the systems need to be monetized for generating revenue. Therefore, pricing and allocation rules need to be set and contracts need to be in place to support the interactions of various different stakeholders. An illustration of the metrics and the associated objectives are provided in Figure 2.3.

2.5   Overview of Scientific Methodologies

The key focus of this book is to design effective mechanisms for MC-IoT applications. Mechanism Design (MD) theory [102] is the engineering arm of economics, which is attributed to Leonid Hurwicz, Eric Maskin, and Roger Myerson, who have received the Nobel Prize in 2007. Although, the original

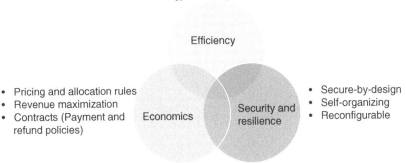

- Quality of service/experience
- Connectivity/reachability
- Energy efficiency

Efficiency

- Pricing and allocation rules
- Revenue maximization
- Contracts (Payment and refund policies)

Economics

Security and resilience

- Secure-by-design
- Self-organizing
- Reconfigurable

Figure 2.3 Key goals and objectives in IoT system design.

application of MD is in the economics context where designing incentives is required to achieve economic outcomes. However, in modern communication systems and the IoT, there is a need to design mechanisms for achieving a diverse range of objectives from system performance to revenue maximization to system security. The fundamental idea behind MD is to design a mechanism that, if implemented and enforced, would eventually lead to the desired outcomes under the dynamic conditions of the IoT systems and networks. The essential idea is to anticipate the **equilibrium** or eventual state of the system. This information can be used to develop a strategy that can steer the equilibrium to the desired state given the system can be controlled. In order for the mechanism to be implementable, it has to be **individually rational** and **incentive compatible**. Individual rationality is also referred to as the participation constraint since the participant in the mechanism must have some positive utility or incentive to be part of the mechanism. Incentive compatibility implies that the participant should not have any incentives to misreport its true type or nature to the mechanism designer. If the mechanism is well designed, the participants would not be pretentious and would participate in an honest manner. This idea is used throughout the dissertation, albeit in a non-traditional sense. Particularly, in the context of IoT and autonomous operation, the role of the mechanism is to devise a set of rules and policies that enable the desired configuration of systems and performance to achieve the objectives laid forth in Section 2.4.

Part II

Design Challenges in MC-IoT

3

Wireless Connectivity Challenges

Wireless connectivity is among the first-mile challenges in the Internet of Things (IoT) ecosystem. In fact, it is a key enabler for the IoT and oftentimes depends on the available communication infrastructure. If traditional wireless communication infrastructure such as base stations of routers are available, it can be used as a vehicle for IoT connectivity. However, in its absence, non-conventional and ground-up techniques can be used to ensure physical wireless connectivity.

3.1 Spectrum Scarcity and Reservation Based Access

Most existing IoT devices utilize traditional wireless personal area network (WPAN) communication technologies [7] such as WiFi, Bluetooth, Zigbee, etc., which are inherently short-range privately administered networks. However, the current focus of network operators is on developing low-power wide-area network (LPWA) technologies [107] dedicated for IoT communication for providing reliable communication and services for large scale IoT systems in smart cities along the same lines as the cellular data networks.

Due to a predicted massive surge in the number of IoT devices in the future owing to widespread adoption of the technology [40, 41], there will be an acute shortage of wireless spectrum for dedicated allocation to these systems. The spectrum requirements for massive IoT networks [44] will exceed the capacity of the unlicensed spectrum bands. Therefore, exploring the sub-GHz spectrum and opportunistically accessing whitespaces in existing licensed spectrum bands will be inevitable to cater for the massive wireless connectivity demand. Several traditional approaches have been developed for mitigating this issue such as new medium access control (MAC) layer schemes [12] to improve the efficiency of multiple access as well as cognitive radio technologies (CRN) [86] for opportunistically accessing licensed spectra. One of the candidate LPWA technologies that have emerged recently is known

Resource Management for On-Demand Mission-Critical Internet of Things Applications, First Edition.
Junaid Farooq and Quanyan Zhu.
© 2021 John Wiley & Sons, Inc. Published 2021 by John Wiley & Sons, Inc.

as ultra-narrow band (UNB) [75], which uses the random frequency and time multiple access (RFTMA) [77] at the MAC layer in which each user selects a time and frequency randomly for transmission. Although it can improve the capacity of the system, there is no performance and reliability guarantees, which may be crucial for mission-critical applications.

Spectrum reservation is an essential approach to provide guarantees for mission-critical (MC) applications. It has been shown that reservation of spectrum leads to efficiency and reliability of communication, which is particularly useful in MC and emergency applications [67]. However, the implementation of spectrum reservation needs to be appropriately incentivized for it to be used in practice. Most works in literature dealing with spectrum reservation are focused mainly on the protocol design aspects such as [68, 69]. Those that deal with the economic perspectives only investigate static contracts to establish the quantity and price of spectrum to be reserved, which assume perfect knowledge of the application requirements [84]. However, in most practical situations particularly emergency networks, the need for spectrum access cannot be perfectly known ahead of time. If the spectrum has not been reserved *a priori*, then the application may have to contend for channel access resulting in significant delays in the communication, which might be costly in MC and emergency scenarios. One of the solutions to this problem is that the network operators offer dynamic contracts to applications whereby an advanced payment is made earlier to reserve the spectrum in the form of a time-frequency (TF) block for future use (Figure 3.1). However, prior to actually using the spectrum, the application may request for a rebate and release the reserved spectrum for use by other applications. Such dynamic contracts can foresee the risks and uncertainties in the future and reduce them through reservation and prioritization (Figure 3.2).

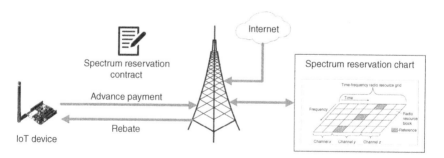

Figure 3.1 A spectrum reservation based UNB-IoT system where the IoT devices only communicate with the AP during the designated time-frequency blocks. The access point (AP) offers a dynamic contract with advance payment and rebate amounts to the devices and maintains a spectrum reservation chart for a fixed time period.

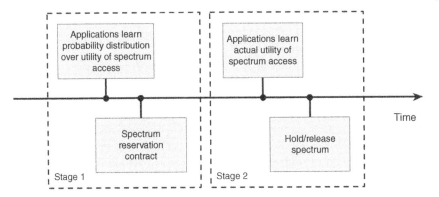

Figure 3.2 Stages of spectrum reservation contract. In the first stage, the applications privately obtain a probability distribution over their utility of the spectrum reservation based on which they opt for one of the contracts by the operator. Once the true utility of the reserved spectrum is known, the applications either hold the spectrum or release it for the agreed rebate amount.

3.2 Connectivity in Remote Environments

Connectivity between smart devices is vital in enabling the emerging paradigm of the IoT [132]. The fundamental goal of the IoT is to inter-connect smart objects so that they can exchange data and leverage the capabilities of each other to achieve individual and/or network objectives such as high situational awareness, efficiency, accuracy, and revenue, etc. This connectivity relies on wireless communication networks that have their limitations based on the communication technologies involved. Existing IoT devices are connected to an access point using WPAN technologies [7] such as WiFi, Bluetooth, Zigbee, etc. The access points are in-turn connected to the wired or wireless backhaul networks using wide area network (WAN) technologies [103]. The backhaul network enables connectivity and accessibility between things that are geographically separated. However, they may not always be available such as in remote areas [33], disaster struck areas [136], and battlefields [39, 41]. UAVs and mobile ground stations are the most viable candidates for providing connectivity in such situations. For instance, during the hurricane *Harvey*, nearly 95% of the cellular sites in Rockport, Texas, went out of service resulting in nearly a complete communication blackout in the region [71]. In such emergency scenarios, where the traditional communication infrastructure is completely devastated, UAVs can prove to be a promising solution to help create a temporary network and resume connectivity in a short span of time. Therefore, there is a growing interest toward the use of drones and UAVs as mobile aerial base stations (BSs) to assist existing cellular LTE networks [65], public safety networks [61], and intelligent transportation systems [93]. While this is promising in urban areas where there is high availability of cellular networks that can be used to connect the UAVs to the backhaul, it might not be possible in rural and/or remote regions.

Due to the absence of traditional communication infrastructure and backhaul networks, the remotely deployed IoT requires a multi-layer architecture comprising of an overlay network of mobile access points (MAPs) to interconnect the spatially dispersed mobile smart devices (MSDs). The MAPs exploit device-to-device (D2D) communications [13, 44] for connecting with other MAPs, while the MSDs connect to one of the available MAPs for communication. The problem in such settings is to efficiently deploy the overlay network that provides coverage to all the MSDs as well as maintaining connectivity between the MAPs. Since the MSDs can be located in spatial clusters that are arbitrarily separated, the MAPs should be deployed in a way that they remain connected, i.e. each MAP is reachable from other MAPs using D2D communications. This requirement makes it a challenging network planning and design problem. Figure 3.3 provides a macroscopic view of one such scenario where the MAPs are appropriately deployed enabling a local inter-network of MSDs without any traditional communication infrastructure. It can be easily connected to the Internet to achieve pervasive connectivity and control over the MSDs. Note that with aerial MAPs, there is an added flexibility to position the BSs

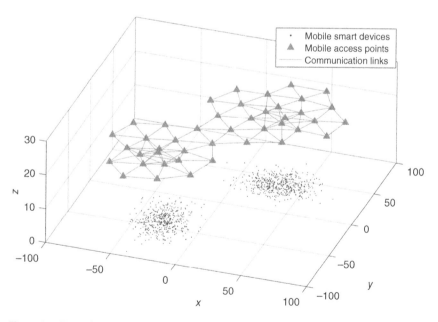

Figure 3.3 Example scenario of spatially clustered mobile smart devices interconnected by an overlay network of mobile access points.

arbitrarily in space, which might not be possible with other traditional types of access points.

There has been an increasing focus recently toward the use of aerial BSs to complement wireless connectivity alongside traditional communication infrastructure [124]. This is due to the significant advances in drone technology coupled with a massive demand for wireless connectivity with the emergence of the IoT. Several works have considered the use of drone BSs to supplement the coverage of existing cellular networks [16]. In [157], the authors have provided a review of the opportunities and challenges in using UAVs for wireless communications. Efficient deployment is undoubtedly one of the key challenges in such communication networks. Therefore, several efforts have been invested in this direction. In [99], the authors use the classical *optimal transport* framework to obtain a power-efficient deployment of UAVs with the objective to collect data from remotely deployed sensors and not to inter-connect BSs in the air. Similarly, in [100], an energy efficient UAV placement strategy is developed for the IoT networks. Other examples of works that use a centralized optimization based approach to the UAV placement problem are [2, 87, 125]. In [65], the authors have proposed a backhaul-aware deployment that is applicable to settings with traditional communication infrastructure. However, in the context of remote IoT, the placement problem is more complex as the BSs have to rely on D2D communication to maintain connectivity. Moreover, in certain applications, it might be inevitable to use such drone-assisted multi-layer architecture.

Most existing works [2, 99, 125] dealing with UAV placement formulate the BS deployment problem as a modification of the well-known *facility location* problem, also referred to as the *p-median* problem [29], from operations research. However, despite being NP-hard, the solution to the facility location problem is not sufficient to ensure the inter-connectivity of the facilities. In our case, the MAPs are wireless devices that have limited communication range and have to be located in sufficient proximity to communicate reliably. Our goal is to place the MAPs in a connected configuration to enable inter-connectivity between the underlying MSDs using D2D links, which are unique to the wireless network setting. This problem is significantly more complex than the multi-facility location problem. Hence, a globally optimal solution to this problem is not easy to obtain. Moreover, a centralized solution is also less attractive due to the practical limitations in the scenario considered in this book since the two-layer network cannot be coordinated by a central planner. Another commonly used approach in the literature [98, 100] utilizes the *circle packing*, also referred to as disk packing, solution that aims to cover the area inside a polygon using non-overlapping circles. Again, it is an NP-hard problem and heuristic algorithms exist only for polygon shapes that cannot be extended to arbitrary boundaries making it less attractive for use in the UAV placement frameworks.

3.3   IoT Networks in Adversarial Environments

IoT devices are manufactured by different vendors without strong regulations on embedding cyber security features in the software [48]. To reduce cost and time-to-market, security issues may be overlooked by device manufacturers [137]. In addition to inherent software vulnerabilities, several other factors increase the risk of cyber attacks on these devices [78, 149]. One of the risks is the use of stock passwords to access the control panel of these devices. Moreover, most IoT devices are left to operate on consumer premises without regular maintenance. It exposes them to the risk of being infected and controlled by malicious software processes, referred to as *malware* [34, 150]. It is also possible that consumers might willingly accept to install certain processes or applications on their devices in return for financial incentives, completely unaware of the fact that they might be used to launch a distributed denial of service (DDoS) attack [73, 83] on the network at a later stage.

Botnets have become a significant threat to computer and communication networks in the last decade [45]. A botnet is a network of devices infected by malicious software and controlled by an external operator referred to as the *botmaster* [143]. Often, the malware infiltrates the network stealthily over time in a self-replicating manner before being instructed by the botmaster to trigger an attack. The objective of the botnet is to cause disruption in service provisioning leading to loss of operation and sometimes with the intent of obtaining ransom [14]. The most famous botnet attack in recent history has been the *Mirai* in 2016 [9]. Recently, researchers have identified variants of the Mirai botnet referred to as the *IoTroop* or *Reaper* that is aimed at using IoT devices to launch DDoS attacks [97]. It is a powerful botnet that comprises of compromised domestic wireless routers, TVs, DVRs, and surveillance cameras exploiting vulnerabilities in devices from major manufacturers.

In the case of wireless IoT networks, the malware may spread from one device to another among devices that are in close geographical proximity [72]. Due to the absence of centralized connectivity, the botmaster is compelled to use the same D2D links to issue control commands for coordinating an attack. Seed viruses may be planted into the networks using malicious or infected IoT devices or even using unmanned aerial vehicles (UAVs) [112]. Moreover, the botmaster may change the malware code dynamically and may issue control commands to launch a wireless denial of service (WDoS) attack [111]. It is different from traditional DDoS attacks as services do not have to be taken off the Internet. Instead, the goal is to exploit medium access control (MAC) vulnerabilities in wireless devices to generate superfluous traffic that sabotages legitimate operation [19]. The D2D nature of the wireless communication network makes it harder to launch a coordinated DDoS. However, at the same time, it is also hard to defend against it as a network of devices contributes to the attack and there is no single source. Therefore, the best strategy for a network defender is to prevent the dynamic development of a large scale botnet and limit its ability to launch a DDoS.

Several dynamic processes might be burgeoning in the network at the same time. Malware in an infected device might be attempting to replicate itself in nearby devices. Furthermore, the infected devices also share control commands with other infected devices to agree on an attack point. On the other hand, the network defense mechanisms are also in place which periodically patch[1] the devices. The patching frequency of devices needs to be carefully selected as it negatively affects the regular device operation. Particularly, if a device acts as a hub, i.e. connecting multiple devices together, the impact of downtime will be much more severe. In order to make such optimal patching frequency decisions, we need a theoretical model that can accurately capture the connectivity characteristics of the network and incorporate the continuing dynamic processes.

While the modeling and analysis of traditional Internet based botnets is also important due to its huge monetary and non-monetary impact, there have been some efforts to prevent and control them. However, the botnets in wireless IoT systems need special attention due to the current lack of awareness and the increased security vulnerability of IoT devices. Despite the impending security threat to a massive number of unprotected IoT devices and systems, there is a severe dearth of systematic methodologies for understanding such systems from a security standpoint.

This necessitates the development of exclusive models for such wireless IoT networks, which can capture the spatial distribution of the devices and the dynamic processes of malware infiltration, control command propagation, and device patching by the defender. In this chapter, we develop the theoretical underpinnings that allow the modeling and analysis of dynamic botnet formation in wireless IoT networks – optimal approach to avoid this intractability while still yielding a plausible solution. Instead of ensuring that the average densities of informed devices $\hat{I}^{(i)}$ exceeds the respective thresholds T_i, $i \in \{m, mn, o\}$, we impose a constraint on the densities of informed devices that possess a degree equal to the average degree of the network. In other words, we ensure that $\hat{I}^{(i)}_{\mathbb{E}[K_i]} \geq T'_i$ for some $T'_i < T_i, i \in \{m, mn, o\}$. It is reasonable because the proportion of devices with the mean degree contribute the most in the average information spreading.

---

1 Throughout this chapter, the term "patch" refers to attempts made by the defender to bring the device to an un-compromised state, e.g. via power cycling, firmware upgrades, etc.

4

Resource and Service Provisioning Challenges

4.1 Efficient Allocation of Cloud Computing Resources

A cloud service provider (CSP) may have several resources such as computing nodes, storage, databases, etc. that can be used remotely by Internet of Things (IoT) applications. Often, these resources are packaged into virtual machine (VM) instances that act as processing units. When the number of requesting applications is large and the available VMs are limited, as envisioned in massive IoT systems, it is important to select which applications are serviced particularly when the allocation is planned for longer periods of time.

The challenges faced by the CSP in allocating the available VMs to requesting applications are twofold. Firstly, the available VMs at the CSP are limited, so it is important to allocate the most computationally intensive tasks to the available VMs in order to maximize the productivity of the IoT client applications and the generated revenue by charging them appropriately. However, the tasks arrive sequentially at the server and the CSP has to decide immediately to allocate a VM to it or to wait for a more valuable task in the immediate future. The challenge lies in the uncertainty about the nature of upcoming requests in the future. A computationally intensive task may not ever request for service while the low complexity tasks are refused service. It leads to an under-utilization of resources resulting in lower productivity and revenue of the CSP. On the other hand, if the VMs are allocated to low complexity tasks, then a high complexity task may request service in the future and has to be refused due to the unavailability of a VM at the cloud server. Therefore, there is a need for a dynamically efficient mechanism for allocating and pricing the VMs that takes these trade-offs into account. Figure 4.1 illustrates the sequential arrival of IoT computation requests at the CSP.

Resource Management for On-Demand Mission-Critical Internet of Things Applications, First Edition.
Junaid Farooq and Quanyan Zhu.

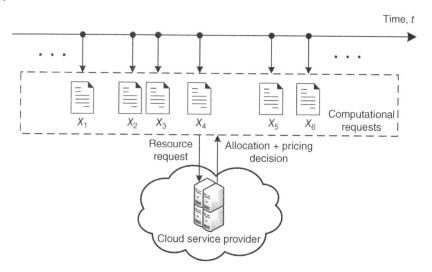

Figure 4.1 Cloud service provider allocating available VMs to sequentially arriving computational requests by IoT client applications.

There has been considerable work in the literature toward resource allocation in cloud computing environments [130]. The focus is mainly on efficient resource management and load balancing for higher availability and performance [91] or resource allocation and pricing for revenue maximization [82, 147]. Regarding dynamic pricing and revenue maximization, several works exist such as [160], which use price control to adjust demand levels. Others have used auction mechanisms to collect bids and allocate available computational resources such as [159]. However, most existing works on resource allocation in cloud computing do not take into account the sequential arrival of computing tasks and the uncertainty about the future. This is essential in the setting of cloud computing because the computational requests are spontaneous and the decision for allocation has to be made immediately upon arrival. A dynamically efficient policy for allocating resources to sequentially arriving agents in order to maximize social welfare was first proposed by Albright [3]. Consequently, a revenue maximizing approach toward sequential allocation of resources has been introduced in [51]. However, their work deals with heterogeneous resources and cannot be used to model situations with identical resources. In cloud-enabled IoT systems, often multiple identical resources such as VMs are available to be allocated to client applications. A framework for pricing the cloud for maximizing revenue is proposed in [147]. However, their solution is based on stochastic dynamic programming, which cannot adapt in real-time scenarios. Our solution provides a dynamically optimal plug and play policy that can be pre-computed and used in real-time using a lookup table.

4.2   Dynamic Pricing in the Cloud

The interconnection of electronic sensors and actuators, known as the IoT [4], is creating enormous opportunities for automating systems around us and improving their efficiency. It is paving the way for the development of smart cities with active monitoring and control of public facilities, smart healthcare, smart transit systems, etc. In recent years, due to the ubiquity of the internet, there has been an increasing trend of offloading computing, control, and storage to the cloud [21]. This is fueling the rapid growth of the IoT as it reduces the physical cost of sensors and actuators. Moreover, connectivity to the cloud opens up endless possibilities for powerful and revolutionary applications, due to the availability of massive computational power and data [128]. Therefore, cloud computing [10] is now becoming an integral part of the IoT ecosystem particularly for applications involving real-time analytics and Big data [20].

There is a wide variety of cloud enabled IoT applications that have different data processing needs. For instance, autonomous vehicles on the roads might require information about the shortest available route to the destination. On the other hand, home users might be controlling appliances remotely based on the information obtained from deployed sensors.

A common characteristic of such applications is the extremely high delay sensitivity. The total delay in the response of a computational request to the cloud server, also referred to as *end-to-end delay*, relies on several factors such as the latency,[1] i.e. round trip time (RTT), in addition to processing time of the computational tasks and the transmission time over the air interface. Figure 4.2

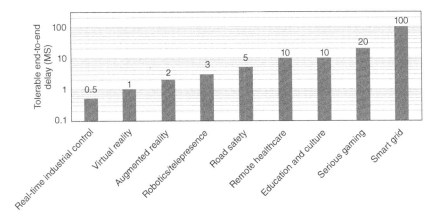

Figure 4.2 Tolerable end-to-end delays of some typical mission-critical IoT applications.

---

1 The term "latency" is used to refer to the delay caused due to the physical distance that is traveled by the data.

shows the maximum tolerable end-to-end delay of some typical mission-critical IoT (MC-IoT) applications [46]. It can be observed that most of the applications require an end-to-end delay of less than 10 ms. Some of them are extremely delay sensitive, requiring an end-to-end delay of 1 ms or less, such as virtual reality (VR) and real-time industrial control applications. Although the existing cloud servers have the computational power to efficiently compute large amounts of data, the location of the server places a bottleneck on the latency. In other words, a certain time delay is inevitable regardless of the size of the task due to the distance the data has to travel before reaching back to its point of origin. Instead of sending the tremendous amounts of data, generated by the IoT to the cloud, it is more efficient if the data is analyzed at the edge of the network, i.e. close to where it is generated, to reduce the latency [153]. Hence, a new computing architecture known as fog computing [47], also referred to as edge computing [129], is now gaining significant attention.

Fog computing is an extension of the cloud such that there are devices located at the edge of the network having computing, storage, and networking capabilities, also referred to as cloudlets or *fog nodes* [15]. Due to the reduced distance, the availability of fog nodes can significantly reduce the response time of cloud server to incoming data [24]. Hence, fog computing is emerging as one of the key enablers of fog-enabled MC-IoT [158] applications. A CSP may have several available fog nodes in addition to the main cloud server for servicing computational requests by MC-IoT applications [138]. An illustration of the hierarchical fog-cloud architecture [126] for an IoT ecosystem is provided in Figure 4.3. Each fog node imposes a certain delay in the response to requests coming from a certain geographical region due to its location. Furthermore, the fog nodes may have multiple virtual machine instances (VMIs), similar to the conventional cloud computing, that process the incoming data. The VMIs have different data processing capabilities according to the allocated computing resources, which result in different processing delays. Altogether, the CSP has a set of available VMIs that are characterized by their overall response time. The MC-IoT applications have a varying level of delay tolerance, i.e. the maximum delay in the response of the cloud, which does not result in degradation of performance in their operation. Once an application requests the cloud for processing a computational task, the job of the CSP is to instantly make a decision of sending the received data to one of the available VMI. Since the requests by MC-IoT applications arrive at random times and have different levels of delay tolerances, the CSP needs to devise a policy according to which the VMIs are allocated and appropriately price them to maximize its revenue.

Since the number available VMIs are limited, it is important to forward the most delay sensitive sets of data to the best[2] available VMI and vice versa to

---

2 The notion of "best" refers to the VMI that offers the lowest end-to-end delay.

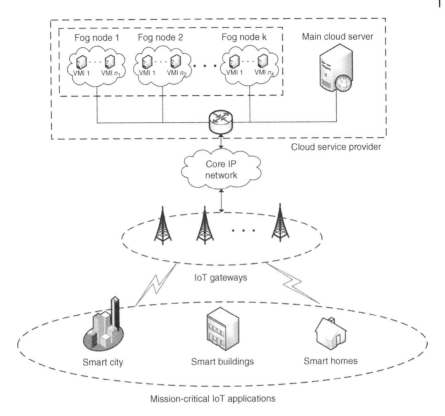

Figure 4.3 Architecture of a fog-enabled IoT ecosystem in which the MC-IoT devices are connected to the CSP via a gateway. The CSP has several fog nodes, equipped with a range of VMIs, in addition to the main cloud server.

deliver a high quality-of-experience (QoE) to the users[3] and consequently generate higher revenue. Note that improving the QoE of client applications enables the CSP to appropriately charge them for premium services. This implies that a highly delay sensitive applications should be allocated to a low end-to-end delay providing VMI and charged higher prices while delay tolerant applications should be allocated longer end-to-end delay providing VMIs at lower prices. However, the challenge lies in the fact that there is limited information about the nature of upcoming requests in the future. A highly delay sensitive application may not request for service while the CSP reserves the best available VMIs, which lead to an inefficient utilization of resources resulting in lower QoE of the users. On the other hand, if the delay

---

3 The term "users" and "clients" are interchangeably used to refer to the client MC-IoT applications that are using the cloud for data processing.

tolerant applications are allocated the best VMIs, then highly delay sensitive applications may request service in the future and may suffer in performance due to the unavailability of VMIs with low end-to-end delay. Secondly, as the time progresses, if the best available VMIs are not utilized while waiting for highly delay sensitive applications, the CSP may lose the opportunity of generating revenues from them at all and it may have been better to allocate them to less delay sensitive applications. Therefore, there is a need for a dynamically efficient policy that takes these tradeoffs into account. We use the following example using the simplest case to elaborate the concept.

Example   Consider a CSP with only a single VMI available for allocation within a time period of 10 hours. Requests arrive sequentially at the CSP according to a Poisson process with an arrival rate of $\lambda = 20$ requests per hour. Each arriving request is assumed to have a random minimum required response rate (exponentially distributed with mean $\alpha = 1$). It implies that there is a high probability of delay tolerant applications arriving while there is a low probability of delay sensitive applications requesting service. The CSP sets a threshold specifying the barrier on the required response rate beyond which the first arriving request will be allocated to the VMI resulting in an efficiency equal to the product of the barrier and the response rate of the VMI. Consequently, a price equivalent to the qualification threshold is charged. The expected revenue of the CSP is then the product of the barrier and the probability of a qualifying request arriving. If the barrier is set too low, the probability of a qualifying request within the allocation period will be high but the revenue generated will be low. Similarly, if the bar is set too high, then the probability of an eligible request will be low but the revenue generated if successful allocation takes place will be high. Figure 4.4 shows the expected revenue against the qualification threshold set by the CSP. It is clear that there exists an optimal qualification threshold that results in maximum revenue of the CSP under uncertainty of arriving requests.

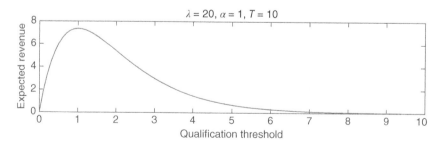

Figure 4.4 Expected revenue of the CSP for a single available VMI with varying qualification threshold.

Note that the example uses a static barrier to filter out computational requests. However, since the allocation period is finite, the threshold needs to be dynamic, i.e. changing with time, in order to maximize the generated revenue. Therefore, there is a need to develop an integrated policy framework that uses the QoE of the users as a basis to maximize the expected revenue generated by the CSP.

Latency is considered as one of the biggest hurdles in the use of cloud computing for real-time and MC applications [76]. Several methods have been used in literature to reduce the response time in remote computing such as shortening the frame size, using instant-access resource allocation to avoid the medium-access contention delays, and moving the network processing feature to the edge of the network. The edge-centric computing [49] is by far the most promising approach to reducing latency. The availability of fog nodes results in an opportunity for the CSP to offer differentiated services in terms of end-to-end delay to cater the demands of MC-IoT and to maximize revenue generation.

Efficient resource management in the for IoT applications is currently an active area of research in the context of cloud computing [30]. Several different objectives have been targeted for workload offloading to the fog-cloud architecture. For instance, in [131], a framework for allocation of combined fog-cloud resources for IoT services is proposed to minimize the latency experienced by the services. On the other hand, workload allocation in fog-cloud computing for balanced delay and power consumption is provided in [31]. Similarly, the works in [151, 152], and [79] provide useful frameworks for resource provisioning in fog based systems for IoT networks. A quality of service (QoS) centric approach has been proposed in [17]. However, these works do not focus on the MC-IoT applications and the revenue maximization aspect.

Several effective resource allocation and pricing strategies have been established for traditional cloud computing environments. For instance, dynamic allocation of resources for spot markets in cloud computing environments has been explored in [160]. However, it aims to optimize the allocation of computational resources according to the price customers are willing to pay but does not cater for the uncertainty in demand. The idea of dynamic pricing and revenue maximization in the presence of stochastic demand has also been investigated in the context of cloud computing [147]. However, it does not use price discrimination for differentiated services offered by the cloud, which is imperative in the case of fog-enabled MC-IoT.

4.3 Spatio-Temporal Urban Service Provisioning

The integrated networks of engineered cyber and physical systems, referred to as the IoT, provide the enabling technology for cities to greatly improve the

security, life, and wellbeing of its citizens [139]. The smart cities paradigm is creating a plethora of opportunities to efficiently utilize available city resources[4]
[118]. Resource allocation problems occur in a wide variety of scenarios in
smart cities such as in disaster management, emergency response systems, public safety systems, taxi pickups, controlling epidemic outbreaks, data collection using wireless sensors, etc. [119]. Typically, there is a centralized *source
node* having a finite number of available resources that need to be allocated to
demand nodes or *service requests* that arrive sequentially over time at random
locations with varying severity or *magnitude*. The task of the source node is
to allocate available resources to service requests in real-time to maximize the
total expected utility[5] obtained from allocation.

Figure 4.5 illustrates a snapshot of the spatially disperse service requests
that have accumulated over time with reference to a centralized source node
placed at the origin serving a circular region. The height of impulses at the
request locations represents the severity or magnitude of the demand. One of
the potential ways to provision resources is to immediately allocate upon the
arrival of a request regardless of its magnitude or distance. However, it may
result in myopic decisions leading to a suboptimal outcome. In many practical
applications, the decision is made once a pool of requests is available. The
challenge in the allocation decision by the source is twofold. Firstly, a very high
magnitude demand request may need service but it may be located at a farther
distance from the source. Secondly, the source needs to decide whether to
allocate a resource to one of the current requests or to wait for future requests,
which may result in higher benefit. While discarding allocation to current

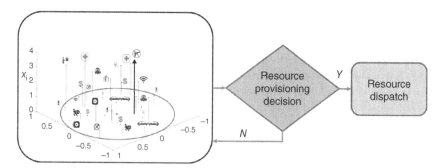

Figure 4.5 Illustration of the centralized resource allocation problem during one time slot.
The impulse height represents the severity or magnitude of requests. The maximum
magnitude request is highlighted by a bold impulse with an arrowhead.

---

4 The term "resources" may generically refer to emergency response units, taxis, wireless
channels, aerial vehicles, vaccines, etc.
5 The term "utility" may refer to different quantities such as social welfare, revenue, etc.,
according to the application scenario.

requests, the source might keep waiting for better requests that never arrive in the future. In essence, the question is where to draw the line for allocation in terms of distance and magnitude of request as well as the waiting time of allocation.

As an example, consider a data collection problem where a centralized base station (BS) schedules the uplink transmission of spatially deployed sensor nodes. The sensor nodes randomly request for data collection to the BS informing it about their location as well as the channel state information.[6] The base station then decides whether to collect data from the sensor or to discard it. If the sensor is close to the BS but the channel gain is extremely low, then the probability of successfully obtaining data from it might be low. However, if the sensor is located far away but the channel gain is extremely high, there might be a strong chance of successfully retrieving data from it. Therefore, the BS needs to optimally select the nodes, from which to collect data, considering the limited number of time slots that are available to obtain data. Note that both the spatial and temporal components are highly important in the decision-making while allocating decisions.

Spatio-temporal service provisioning has been investigated in different contexts and scenarios in the literature. The most common application is the dispatch of policing services [36] or emergency response units, e.g. ambulances [62] to crime incidents and accidents respectively in cities [156]. In [54], a heuristic based algorithmic approach is used to dispatching policing resources to citizen's calls according to the severity of the incidents. Similarly, in [163], police units are dispatched in response to crime incidents using graph mapping and routing algorithms. However, these do not consider the temporal aspects of the allocation as well as the limitation on the number of available resources in comparison with the requests. In other words, the allocation decision is not based on consideration of the future requests.

Another application area spatio-temporal resource provisioning is the dispatch of taxis to pick up requests [148] in a geographical region. In [123] and [88], algorithms have been proposed to match customers with taxis to minimize the waiting time for customers as well as the idle time for taxis. However, these algorithms are based on static assignment solutions that do not consider the dynamics of the requests making the solutions less attractive. Similarly, several frameworks have also been developed to tackle the problem of resource provisioning in other scenarios, e.g. spatial crowdsourcing [55], task scheduling in cloud data centers [22, 85, 127] planning of distributed energy resources [144], urban sensing [116, 162], smart data collection [161], etc. However, these works mainly use static assignment and are based on heuristic approaches to optimize the resource provisioning problem. Those

---

6 It is assumed that sensor nodes are always able to communicate low bandwidth signaling information to the BS via separate control channels.

that use arrival dynamics, e.g. in a crowd sensing application [59], make use of queueing models, in which the unserviced requests remain in the system. Some work has also been done in the distribution of vaccines for controlling epidemic outbreaks [81]. In such cases, the epidemic evolves according to its underlying dynamics while the goal is to select the time and location for distributing vaccines. This problem is fundamentally different as the demand is not perishable, but in fact evolves with time. However, in this section, we deal with spontaneous and perishable requests distributed in the spatial and temporal domains.

The real-time allocation of resources to stochastic incoming requests has been developed in the context of cloud computing [40]. However, it does not consider the spatial dimension of requests and there is no waiting time involved. In other words, an allocation decision is made immediately upon arrival of requests. In practical applications, the decision-making process may not be immediate. Hence, the source node may wait for an allocation period and decide to allocate a resource that may lead to the maximum benefit. In general, there is a lack of systematic and provably optimal approaches for centralized resource allocation to spatio-temporal service requests.

Part III

Wireless Connectivity Mechanisms for MC-IoT

5

Reservation-Based Spectrum Access Contracts

Spectrum reservation is emerging as one of the potential solutions to cater for the communication needs of massive number of wireless Internet of Things (IoT) devices with reliability constraints particularly in mission-critical scenarios. In most mission-critical systems, the true utility of a reservation may not be completely known ahead of time as the unforeseen events might not be completely predictable. In this section, we present a dynamic contract approach where an advance payment is made at the time of reservation based on partial information about spectrum reservation utility. Once the complete information is obtained, a rebate on the payment is made if the reservation is released. A contract theoretic approach is presented to design an incentivized mechanism that coerces the applications to reveal their true application type resulting in greater profitability of the IoT network operator. The operator offers a menu of contracts with advanced payments and rebate to the IoT applications without having knowledge about the types of applications. The decision of the applications in selecting a contract leads to a revelation of their true type to the operator, which allows it to generate higher profits than a traditional spectrum auction mechanism. Under some assumptions on distribution of the utility of the applications, closed form solutions for the optimal dynamic spectrum reservation contract are provided and the sensitivity against system parameters is analyzed.

5.1 Reservation of Time–Frequency Blocks in the Spectrum

In this section, we first describe the network model comprising of details regarding the availability and cost of obtaining an idle time-frequency (TF) block. Then, a description of utility achieved by applications from reserving the spectrum in advance is discussed.

Resource Management for On-Demand Mission-Critical Internet of Things Applications, First Edition.
Junaid Farooq and Quanyan Zhu.
© 2021 John Wiley & Sons, Inc. Published 2021 by John Wiley & Sons, Inc.

5.1.1 Network Model

We assume that a single IoT operator is coordinating the communication between low power IoT devices using ultra narrow band (UNB) transmissions. We consider a single access point (AP) serving a wide area network of IoT devices. The AP reserves TF blocks in the available whitespace in existing licensed spectra for a fixed duration T in the future. Let n_t denote the number of available channels of equal bandwidth β at time t. An illustration of the spectrum resources available to the AP for the duration T is provided in Figure 5.1. At each time $t = 1, \ldots, T$, a random number of channels is available due to the uncontrolled activity of the primary users in licensed bands. Let $\kappa_t = g(n_t)$, where $g : \mathbb{Z}^+ \to \mathbb{R}$, represents the cost of obtaining a channel at time t that is related to the number of available channels. We denote the time average of the channel cost over the duration T by $\kappa = \frac{1}{T}\sum_{t=1}^{T} \kappa_t$.

The IoT applications are categorized into two broad types, i.e. MC or non-MC. Let $\pi_c \in [0, 1]$ represent the proportion of MC applications and $\pi_n = 1 - \pi_c$ be the proportion of non-MC IoT applications. The AP enforces a reservation based access scheme in which the applications reserve TF blocks for the duration T in advance. If before the time of channel access, the application determines that the transmission is not needed, it can release the TF block in advance and request for a partial rebate on the initial payment.[1] Hence, this transaction can be summarized into two stages. At the beginning of the first stage, the applications privately make an assessment of their transmission requirements in the future. Since emergency situations and/or other unforeseen events cannot be completely predicted, the applications obtain a probability distribution over their utility of reserving the channel. At the end of the first stage, the service provider and the applications enter into a contract that includes an advance payment and a rebate agreement. At the beginning of the second period, the applications learn their true utility of the

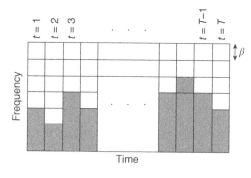

Figure 5.1 Available spectrum divided into time–frequency blocks for reservation. The shaded regions indicate blocks occupied by the primary user.

---

1 The payments can be realized in several ways such using credits where each device has a fixed number of credits to use for spectrum access at the beginning of the first stage.

reserved spectrum. If the utility is higher than the rebated amount in monetary terms, the application holds the channel. Otherwise, it prefers to release the channel and obtain a rebate at the end of the second period.

For each time instant, $t = 1, \ldots, T$, two contracts are offered by the network operator based on the average cost of obtaining an idle channel. The proportion of MC and non-MC applications vary during each time slot due to the relative frequency of transmissions of each type of application. For instance, if an MC application has to transmit at each time slot, then the number of contending MC applications will be the same for all time slots. Similarly, if the non-MC applications have to transmit once after 10 time slots, then the non-MC applications will only contribute in the proportion once after every 10 time slots.

5.1.2   Utility of Spectrum Reservation

In the absence of spectrum reservation, the IoT devices have to contend for spectrum access by employing a multiple access protocol involving some randomness in which case, the average delay in channel access or the average number of transmission until successful message delivery is denoted by δ. It is natural to associate the spectrum reservation utility of the applications to the expected channel access delay experienced if there was no reserved spectrum. The average access delay is a function of the number of simultaneously transmitting devices, their physical placement relative to other neighboring devices, and the multiple access protocol used. Assuming that δ represents the waiting time until successful spectrum access, we model it as an exponential random variable with mean $\frac{1}{\lambda}$. Let the utility of an application for reserving a TF block in the future be represented by $V_\chi = h_\chi(\delta)$, $\chi \in \{c, n\}$ referring to the MC and non-MC applications, respectively. $h_\chi(\cdot)$ is a transformation of the average delay into a monetary value. We assume that the transformation $h(\cdot)$ does not change the distribution of the delay except the mean and the variance. Therefore, we can claim that V_χ is also exponentially distributed with intensity $\frac{1}{\lambda_\chi}$, $\chi \in \{c, n\}$. Consequently, we can state that $f_{V_\chi}(v) = \lambda_\chi e^{-\lambda_\chi v}$ and $F_{V_\chi}(v) = 1 - e^{-\lambda_\chi v}$. With a slight abuse of notation, we will refer to these functions as simply $f_\chi(v)$ and $F_\chi(v)$, $\chi \in \{c, n\}$ in the sequel. Furthermore, we assume that $\lambda_c \leq \lambda_n$, which implies that the MC applications have a higher utility of reserving the channel as compared to the non-MC applications. This is a reasonable assumption as, inherently, MC applications are more sensitive to the average delay in spectrum access.

5.2   Dynamic Contract Formulation

In this section, we first describe the objective of the IoT service provider and then provide a detailed description of the methodology used to achieve the desired objective.

5.2.1 Objective of Network Operator

The objective of the IoT network operator is to create a menu of contracts for the two main types of IoT applications, which allow it to coerce them into revealing their private information and using it to generate higher profits. The challenge lies in the information asymmetry between the applications and the service provider. The applications know more about their utility of a TF block than the service provider. If the service provider had information about the type of application requesting the TF blocks, it can use price discrimination to extract maximum profit by charging higher prices to MC applications. However, in the absence of information about the applications, the service provider has to design a mechanism that results in the applications revealing their true types by selecting the contract that suits them. This mechanism is in the form of a refund contract where the applications learn their utilities sequentially.

In summary, the objective of the operator is to optimally design the advance payments and rebates for the MC and non-MC applications represented by the tuple $C = \{p_c, r_c, p_n, r_n\}$, where p_c and p_n are the advance payments, respectively, at the end of the first stage while r_c and r_n is the rebate offered by the network operator in case the applications release the reserved TF blocks at the end of the second stage. The IoT applications will select one of the contracts and if the mechanism is well designed, the applications will not have any incentive to deviate from their true preferences. Hence, in this process, they reveal their true types to the operator leading to higher profitability.

5.2.2 Spectrum Reservation Contract

The contract is established at the end of the first stage while only a probability distribution over the applications' valuation is known to them privately.

5.2.2.1 Operator Profitability

Given the contract tuple $C = \{p_c, r_c, p_n, r_n\}$, the expected profit of the IoT network operator can be expressed as follows:

$$\Pi(C) = \sum_{\chi \in \{c,n\}} \pi_\chi (p_\chi - r_\chi F_\chi(r_\chi) - \kappa(1 - F_\chi(r_\chi))),$$

$$= \sum_{\chi \in \{c,n\}} \pi_\chi \left(p_\chi - r_\chi + e^{-\lambda_\chi r_\chi}(r_\chi - \kappa) \right). \tag{5.1}$$

Note that p_χ is the advance payment received by the network operator, $r_\chi F_\chi(r_\chi)$ is the expected amount of rebate paid back to the application, and $\kappa(1 - F_\chi(r_\chi))$ is the expected cost of obtaining an idle channel by the operator. The expected profit from both MC and non-MC applications is added after being weighted by their proportions in the network.

5.2.2.2 IC and IR Constraints

In order for the spectrum reservation mechanism to be implementable, it has to be individually rational (IR), i.e. both the MC and non-MC applications will only participate in the spectrum reservation if at best on average, they do not receive a lower utility than the payment they make. Note that at the end of the second stage, the application will always prefer to release the channel in return for a rebate if the amount refunded is greater that its actual utility of the channel. Therefore, the expected return of the application can be written as $\int_0^\infty R_\chi(v) f_\chi(v) dv$, where $R_\chi(v) = \max(r_\chi, v)$. Hence, the expected utility of the application for reserving the channel can be expressed as $r_\chi F_\chi(r_\chi) + \int_{r_\chi}^\infty x f_X(x) dx$ and the IR implies that it should be at least as much as the payment made for reservation. This can be expressed as follows:

$$r_\chi F_\chi(r_\chi) + \int_{r_\chi}^\infty x f_X(x) dx \geq p_\chi, \quad \chi \in \{c, n\}. \tag{5.2}$$

Moreover, the mechanism also needs to be incentive compatible (IC), i.e. there is no incentive for any type of application to hide their true type from the operator by choosing a different contract. In other words, the MC applications should be better off choosing the MC contract and vice versa. For the two-type case, this can be formally expressed by the following two constraints:

$$r_c F_c(r_c) + \int_{r_c}^\infty v f_c(v) dv - p_c \geq r_n F_c(r_n) + \int_{r_n}^\infty v f_c(v) dv - p_n, \tag{5.3}$$

$$r_n F_n(r_n) + \int_{r_n}^\infty v f_n(v) dv - p_n \geq r_c F_n(r_c) + \int_{r_c}^\infty v f_n(v) dv - p_c. \tag{5.4}$$

The next step is combine these set of equations for the case considered to write the optimal contracting problem.

5.2.3 Optimal Contracting Problem

The optimal contracting problem can then be formally written as follows:

$$\max_C \quad \Pi(C) = \sum_{\chi \in \{c,n\}} \pi_\chi \left(p_\chi - r_\chi + e^{-\lambda_\chi r_\chi} (r_\chi - \kappa) \right), \tag{5.5}$$

subject to

$$(\text{IR}_\chi) \qquad r_\chi + \frac{1}{\lambda_\chi} e^{-\lambda_\chi r_\chi} \geq p_\chi, \ \chi = \{c, n\}, \tag{5.6}$$

$$(\text{IC}_{c,n}) \qquad r_c + \frac{e^{-\lambda_c r_c}}{\lambda_c} - p_c \geq r_n + \frac{e^{-\lambda_c r_n}}{\lambda_c} - p_n, \tag{5.7}$$

$$(\text{IC}_{n,c}) \qquad r_n + \frac{e^{-\lambda_n r_n}}{\lambda_n} - p_n \geq r_c + \frac{e^{-\lambda_n r_c}}{\lambda_n} - p_c. \tag{5.8}$$

The solution to this problem results in the optimal contract C that leads to the maximum profit for the network operator in the scenario where the MC applications learn their utility of channel reservation over time in multiple stages.

5.2.4 Solution to the Optimization Problem

Before we proceed toward the solution to the problem, we first exploit the structure of the constraints to remove the redundancies leading to a simplification of the original problem. The special structure emerges due to the properties of the exponentially distributed utilities of the applications. We describe the notion of first-order stochastic dominance (FSD) in the following lemma.

Lemma 5.1 *The distribution of a random variable $X \in \mathcal{X}$ first order stochastically dominates the distribution of a random variable $Y \in \mathcal{X}$ if $F_X(x) \leq F_Y(x), \forall x \in \mathcal{X}$.*

According to the lemma, it is clear that the distribution of utility of MC applications $F_c(v)$ first order stochastically dominates the distribution of non-MC applications $F_n(v)$ since $\lambda_c \leq \lambda_n$. The implication of this property on the original optimization problem described by Eqs. (5.5) to (5.8) is provided by the following proposition.

Proposition 5.1 *Under the FSD of the distribution of utility of MC applications, the constraints $IC_{c,n}$, and IR_n imply IR_c. Hence, the constraint IR_c is redundant and can be removed from the problem.*

Proof: The constraints $IC_{c,n}$ and IR_n are expressed as follows:

$$r_c + \frac{e^{-\lambda_c r_c}}{\lambda_c} - p_c \geq r_n + \frac{e^{-\lambda_c r_n}}{\lambda_c} - p_n, \tag{5.9}$$

$$r_n + \frac{1}{\lambda_n} e^{-\lambda_n r_n} - p_n \geq 0. \tag{5.10}$$

Since $\lambda_c \leq \lambda_n$, it is clear that $r_n + \frac{e^{-\lambda_c r_n}}{\lambda_c} - p_n \geq r_n + \frac{1}{\lambda_n} e^{-\lambda_n r_n} - p_n \geq 0$. This in turn implies that $r_c + \frac{e^{-\lambda_c r_c}}{\lambda_c} - p_c \geq 0$, which precisely describes the constraint IR_c. Hence, we can remove it from the original problem. $\qquad\square$

Furthermore, note that in the optimal contract, the constraint IR_n is binding, i.e. satisfied with equality. If it does not bind, then increasing $p_\chi, \chi \in \{c, n\}$ equally can lead to an increase in the profit of the IoT network operator. Similarly, the constraint $IC_{c,n}$ also binds since otherwise increasing p_c can lead to an increase in profit. Therefore, we can substitute the constraints IR_n and

$IC_{c,n}$ into the objective function and ignore $IC_{n,c}$ to obtain a relaxed problem as follows:

$$\max_{r_c, r_n} \quad e^{-\lambda_n r_n}\left(\frac{1}{\lambda_n} + \pi_n(r_n - \kappa)\right) - e^{-\lambda_c r_n}\left(\frac{\pi_c}{\lambda_c}\right)$$
$$+ \pi_c\left(e^{-\lambda_c r_c}\left(\frac{1}{\lambda_c} + r_c - \kappa\right)\right). \tag{5.11}$$

Fortunately, the relaxed problem is separable in the decision variables r_c and r_n. Hence, the optimal solution to the relaxed problem can be expressed by the following lemma.

Lemma 5.2 *In the optimal contract offered to the IoT applications, the rebate to the MC applications that maximizes the expected profit equals the average cost of the channel, i.e. $r_c^* = \kappa$. The optimal rebate for the non-MC applications can be obtained by solving the following fixed-point equation:*

$$r_n^* = \kappa + \frac{\pi_c(e^{r_n^*(\lambda_n - \lambda_c)} - 1)}{\pi_n \lambda_n} \tag{5.12}$$

Proof: Let $J(r_c) = \pi_c\left(e^{-\lambda_c r_c}\left(\frac{1}{\lambda_c} + r_c - \kappa\right)\right)$. Since $\frac{dJ(r_c)}{dr_c} = -\lambda_c(r_c - \kappa)$, the optimal amount of rebate offered to the MC applications $r_c^* = \kappa$ and it is indeed a maximizer as $\frac{d^2 J(r_c)}{dr_c^2} = -\lambda_c < 0$. Similarly, let $H(r_n) = e^{-\lambda_n r_n}\left(\frac{\pi_c + \pi_n}{\lambda_n} + \pi_n(r_n - \kappa)\right) - e^{-\lambda_c r_n}\left(\frac{\pi_c}{\lambda_c}\right)$. Now, $\frac{dH(r_n)}{dr_n} = e^{-\lambda_n r_n}(-\pi_c - \lambda_n \pi_n(r_n - \kappa)) + \pi_c e^{-\lambda_c r_c}$. Setting this to zero results in the fixed-point equation given by (5.12). $\square$

Lemma 5.3 *A unique fixed-point solution exists for the optimal rebate for non-MC applications given by (5.12) only if $\pi_c \leq \frac{\lambda_n}{2\lambda_n - \lambda_c}$ and is expressed as follows:*

$$r_n^* = \log\left(\frac{\lambda_n \pi_n}{\pi_c(\lambda_n - \lambda_c)}\right). \tag{5.13}$$

Proof: Let $L(r_n) = \kappa + \frac{\pi_c(e^{r_n(\lambda_n - \lambda_c)} - 1)}{\pi_n \lambda_n}$. Since $L(r_n)$ is an exponentially increasing function of r_n, so a unique fixed-point only exists if $\frac{dL(r_n^*)}{dr_n} = 1$. Solving this results in the expression for optimal rebate for non-MC applications given by the lemma. Furthermore, a valid solution only exists if $r_n \geq 0$ which implies that $\frac{\lambda_n \pi_n}{\pi_c(\lambda_n - \lambda_c)} \geq 1$ resulting in the condition provided for π_c in the lemma. $\square$

Now, for optimality of r_c^* and r_n^*, it is sufficient to show that the solution to the relaxed problem satisfies the $IC_{n,c}$ constraint as well. Since the $IC_{c,n}$ constraints bind with equality, we know that

$$p_n - p_c = r_n - r_c + \frac{e^{-\lambda_c r_n}}{\lambda_c} - \frac{e^{-\lambda_c r_c}}{\lambda_c}. \tag{5.14}$$

In order for $IC_{n,c}$ constraint to be specified, we need to show the following:

$$r_n - r_c - (p_n - p_c) + \frac{e^{-\lambda_n r_n}}{\lambda_n} - \frac{e^{-\lambda_n r_c}}{\lambda_n} \geq 0. \tag{5.15}$$

Substituting (5.14) into (5.15) results in the following:

$$\frac{e^{-\lambda_n r_n}}{\lambda_n} - \frac{e^{-\lambda_n r_c}}{\lambda_n} + \left(\frac{e^{-\lambda_c r_c}}{\lambda_c} - \frac{e^{-\lambda_c r_n}}{\lambda_c} \right) \geq 0, \tag{5.16}$$

which can be shown to be true for $r_c^* = \kappa$ and r_n^* given by Lemma 5.3 using the Taylor series expansion. Once the optimal rebates have been obtained, the optimal advance payment can be obtained using the binding constraints IR_n and $IC_{c,n}$ as expressed by the following corollary.

Corollary 5.1 *In the optimal contract offered to the IoT applications, the advance payment for reserving a TF block by MC and non-MC applications that maximizes the expected profit of the operator can be obtained as follows:*

$$p_n^* = r_n^* + \frac{1}{\lambda_n} e^{-\lambda_n r_n^*}, \tag{5.17}$$

$$p_c^* = p_n^* + r_c^* - r_n^* + \frac{1}{\lambda_c}(e^{-\lambda_c r_c^*} - e^{-\lambda_c r_n^*}). \tag{5.18}$$

This completes the solution to the optimal contract design for spectrum reservation in IoT systems.

5.3 Mission-Oriented Pricing and Refund Policies

In this section, we present the results obtained from analyzing the designed optimal spectrum reservation menu of contracts for varying system parameters. We first describe the network setup and the assumed system parameters. We assume a single monopoly IoT network operator and consider the scenario where an AP coordinates the communication between various IoT devices and the Internet. The AP enforces a reservation based spectrum access system where the applications can only communicate with the AP in a particular TF block if they have a prior reservation. Furthermore, the AP does not know about the nature of the application using the spectrum. Therefore, it only offers two contracts to the applications based on prior information about the proportions of the application types and the distribution of their spectrum reservation utility.

The system parameters are selected as follows: We assume that 20% of the applications connected to the AP are MC and the rest are non-MC, i.e. $\pi_c = 0.2$ and $\pi_n = 0.8$ unless otherwise stated. The time average of the cost of obtaining an idle channel to the network operator is considered to be $\kappa = 0.1$ monetary units (MUs). Unless otherwise stated, we will use $\lambda_c = 0.2\,\mathrm{MU}^{-1}$ and $\lambda_n = 1\,\mathrm{MU}^{-1}$ implying that the expected utility of the MC applications, i.e. 5 MU, is higher than the non-MC applications, i.e. 1 MU before the true utility is learnt. Note that the parameter selection is made for illustrative purposes and does not affect the generality of the results. Using the selected parameters, the optimal amount of rebate offered to the applications is obtained using Lemmas 5.2 and 5.3. Consequently, the optimal advance payments can be obtained using results in Corollary 5.1. The main observations are described in the sequel.

In Figure 5.2, we plot the optimal menu of contracts against varying proportion of MC applications in the network. It can be observed that the optimal amount of rebates (shown by dotted lines) are always lower than the

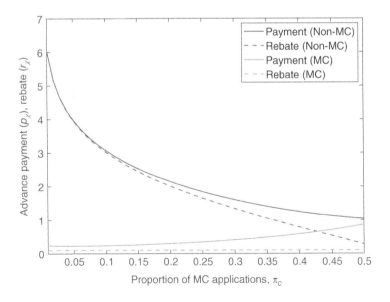

Figure 5.2 Effect of varying proportion of MC applications in the network.

corresponding advance payments (shown by solid lines). Moreover, that the rebate to MC applications in the optimal contract is always equal to the average cost of obtaining a channel, i.e. $\kappa = 0.1$. It is observed that as the proportion of MC applications increases, the advance payment for MC applications in the optimal contract reduces while the optimal rebate for MC applications is fixed at κ. On the other hand, the proportion of non-MC applications decreases implicitly and this also results in a decrease in the advance payment and the rebate. Note that as the proportion of MC applications increase to 0.5, i.e. the MC and non-MC applications are equally distributed, the return profit margin (difference between advance payment and rebate) of the operator becomes equal for both menus.

Figure 5.3 investigates the effect of the average utility of the MC applications prior to contracting on the optimal contracts. Note that increasing λ_c implies that the expected utility of the MC applications decreases. We fix $\lambda_n = 1 \, MU^{-1}$ and sweep λ_c to $\lambda_c = 0.5$. Note that as λ_c increases, i.e. the expected utility decreases, the payment and rebate amounts increase in general to make up for the revenue. Furthermore, the increase is proportional to the composition of the application types.

Finally, in Figure 5.4, we investigate the effect of varying the average cost of opportunistically obtaining an idle TF block in the licensed spectrum. It is intuitive that the advance payments and the rebates for MC applications increase with the increasing cost. However, it is interesting to note that the

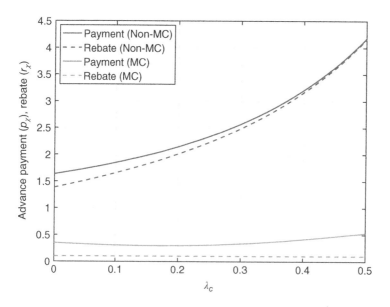

Figure 5.3 Effect of decreasing average valuation of MC applications.

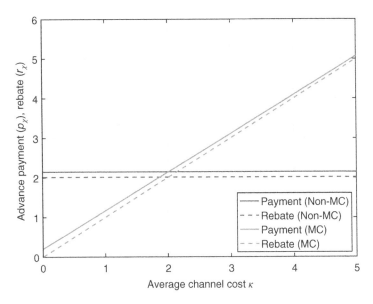

Figure 5.4 Effect of average channel cost on the reservation contract.

rebates increase linearly with the cost while the advance payments increase non-linearly. As the cost becomes significantly high, the rebate policy in the optimal dynamic contract effectively becomes a full refund policy. In other words, the advance payment becomes high on the agreement that a full refund will be issued in case of cancellation later. Furthermore, the advance payments and the rebates designed for non-MC applications are independent of the cost of obtaining the channel.

5.4 Summary and Conclusion

In this chapter, we present a dynamic mechanism design framework to establish a spectrum reservation contract for mission-critical IoT devices in the licensed spectrum using UNB technology where TF blocks can be reserved ahead of time. If the channel is not required later after the initial reservation, the service provider pays back a part of the reservation payment as a rebate. We develop optimal dynamic contracts where the applications learn the reservation utility sequentially over time. The optimal advanced payments and the rebate amounts are obtained in closed form that maximize the expected profit of the network operator. Finally, the behavior and properties of the contract terms are analyzed against different system parameters.

In the presented framework, our design and analysis restricted to use a broad classification of applications, i.e. either MC or non-MC types. In practice, there can be several categories of applications that have different levels of sensitivities to spectrum access delay. Therefore, future extensions in this direction can further enhance the developed model to a continuum of application types covering various levels of mission criticality, which may offer higher price discrimination and consequently higher profitability for the operators.

6

Resilient Connectivity of IoT Using Aerial Networks

6.1 Connectivity in the Absence of Backhaul Networks

The coordination, planning, and design of overlay networks constrained by the underlay devices is a challenging problem. Existing frameworks for placement of unmanned aerial vehicles (UAVs) do not consider the lack of backhaul connectivity and the need for device-to-device (D2D) communication. Furthermore, they ignore the dynamical aspects of connectivity in such networks, which present additional challenges. For instance, the connectivity of devices can be affected by changes in the network, e.g. the mobility of underlay devices or unavailability of overlay devices due to failure or adversarial attacks. To this end, this chapter proposes a feedback based adaptive, self-configurable, and resilient framework for the overlay network that cognitively adapts to the changes in the network to provide reliable connectivity between spatially dispersed smart devices. Results show that the proposed framework requires a significantly lower number of aerial base stations (BSs) to provide higher coverage and connectivity to remotely deployed mobile devices as compared to existing approaches.

Although the existing works in the literature provide optimization based approaches to the UAV placement problem, we believe that this problem is dynamic in nature and hence a more holistic approach is required to obtain an efficient placement of the mobile access points (MAPs) in real-time. In addition to effective initial deployment of MAPs, there is a need for an autonomic, self-organizing, and self-healing overlay network that can continuously adapt and reconfigure according to the constantly changing network conditions [23]. The mobile smartx devices (MSDs) can be highly mobile such as smart handheld devices, wireless sensors, and wearable devices whose mobility can be either individual or collective based on the objective such as a rescue operation or a battlefield mission. Furthermore, the network is also vulnerable to failures and cyber–physical adversarial attacks. Therefore, a distributed and dynamic

Resource Management for On-Demand Mission-Critical Internet of Things Applications, First Edition.
Junaid Farooq and Quanyan Zhu.
© 2021 John Wiley & Sons, Inc. Published 2021 by John Wiley & Sons, Inc.

approach to providing resilient connectivity is essential to cope with the growing scale of the networks toward a massive Internet of Things (IoT) [44]. A large body of work is available in the robotics literature dealing with the coverage of a region by autonomous multi-agent systems [122]. However, they deal primarily with single layer problems and thus the results do not apply directly to a multi-layer network in which one layer aims to provide wireless connectivity to the other. To this end, we propose a feedback based distributed cognitive framework that maintains connectivity of the network and is resilient to the mobility of MSDs and/or failures of the MAPs. The continuous feedback enables the framework to actively react to network changes and appropriately reconfigure the network in response to a failure event that has resulted in loss of connectivity. Simulation results demonstrate that if sufficient MAPs are available, they can be arranged into a desired configuration from arbitrary initial positions and the configuration continuously adapts according to the movement of the MSDs as well as recovers connectivity under varying levels of a random MAP failure event.

6.2 Aerial Base Station Modeling

We consider a finite set of MSDs arbitrarily placed in $\mathbb{R}^2$ denoted by $\mathcal{M} = \{1, \ldots, M\}$ and a finite set of MAPs denoted by $\mathcal{L} = \{1, \ldots, L\}$, placed in $\mathbb{R}^2$ at an elevation of $h \in \mathbb{R}$, for providing connectivity to the MSDs.[1] The Cartesian coordinates of the MSDs at time t are denoted by $\mathbf{y}(t) = [y_1(t), y_2(t), \ldots, y_M(t)]^T$, where $y_i(t) \in \mathbb{R}^2, \forall i \in \mathcal{M}, t \geq 0$. Similarly, the Cartesian coordinates of the MAPs at time t are denoted by $\mathbf{q}(t) = [q_1(t), q_2(t), \ldots, q_L(t)]^T$, where $q_i \in \mathbb{R}^2, \forall i \in \mathcal{L}, t \geq 0$. For brevity of notation, we drop the time index henceforth and assume that the time dependence is implicitly implied. Initially, the MSDs are partitioned into $K \in \mathbb{Z}^+$ sets denoted by $S = \{S_1, S_2, \ldots, S_K\}$. The centroid of each set or cluster is denoted by $\overline{C}_i$. The MAPs have a maximum communication range of $r \in \mathbb{R}^+$, i.e. any two MAPs can communicate only if the Euclidean distance between them is less than r.[2] The communication neighbors of each MAP is represented by the set $\mathcal{N}_i = \{j \in \mathcal{L}, j \neq i : \| q_i - q_j \| \leq r\}, \forall i \in \mathcal{L}$. The quality or strength of the communication links between the MAPs is modeled using a distance dependent decaying function $\alpha_{\{z_1, z_0\}}(z) \in [0, 1]$ with finite cut-offs,[3] expressed

---

1 The MAPs are assumed to have a constant elevation from the ground for simplicity, however, the methodology and results can be readily generalized to varying elevations.
2 The average communication range can be determined using metrics such as the probability of transmission success together with path loss models developed for air to ground communication channels such as in [5, 6].
3 This type of function is referred to as a bump function in Robotics literature. It is pertinent to mention that the cutoff values are crucial while the actual form of the bump function is not important in our application.

as follows [115]:

$$\alpha_{\{z_1,z_0\}}(z) = \begin{cases} 1, & \text{if } 0 \leq z < z_1, \\ 0.5\left(1 + \cos\left(\pi\frac{z-z_1}{z_0-z_1}\right)\right), & \text{if } z_1 \leq z < z_0, \\ 0, & \text{if } z \geq z_0, \end{cases} \tag{6.1}$$

where z_0 and z_1 are constants that define the cut-off values corresponding to 0 and 1, respectively. An illustration of the function with $z_1 = 0.2$ and $z_0 = 1$ is shown in Figure 6.1.

Note that the function $\alpha_{\{z_1,z_0\}}(z) \in [0,1]$, parameterized by z_0 and z_1, represents a generic measure of the communication quality between the MAPs with 1 referring to perfect communication and 0 referring to completely absent communication. However, in practice, the constants z_0 and z_1 need to be carefully selected according to the communication quality requirements based on the physical signal propagation model. One of the possibilities is to use the signal-to-noise-plus-interference-ratio (SINR) to determine the probability of successful transmission between the devices. Thresholds can then be imposed on the coverage probability to obtain the reference distances, which can in turn be normalized to obtain z_0 and z_1. It is pertinent to mention that obtaining the parameters via an accurate characterization of the success probability based on the SINR results also takes into account the effect of interference in the

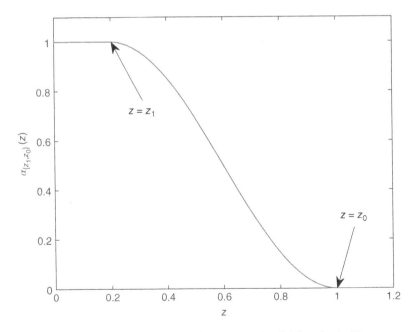

Figure 6.1 Illustration of the strength of communication link function in (1).

network. Several analytical approaches such as stochastic geometry can be used to obtain such a characterization [42, 43]. Furthermore, in order to make the norm measure of a vector differentiable at the origin, a new mapping of the $\mathcal{L}_2$ norm is defined following [105], referred to as the σ-norm:[4]

$$\| x \|_\sigma = \frac{1}{\epsilon}(\sqrt{1 + \epsilon \| x \|^2} - 1), \tag{6.2}$$

where $\epsilon > 0$ is a constant. The smooth adjacency matrix containing the strength of linkages between the MAPs, denoted by $\mathbf{A} = [a_{ij}] \in \mathbb{R}^{L \times L}$, can then be obtained as follows:

$$a_{ij} = \begin{cases} \alpha_{\{\gamma, 1\}} \left(\frac{\|q_i - q_j\|_\sigma}{\|r\|_\sigma} \right), & \text{if } i \neq j, \\ 0, & \text{if } i = j. \end{cases} \tag{6.3}$$

Note that the function in (6.3) uses $z_0 = 1$ and z_1 is replaced by an arbitrary parameter $\gamma \in [0, 1]$ referring to the upper cutoff value. The degree matrix of the MAPs is defined by $\mathbf{D} = [d_{ij}] \in \mathbb{R}^{L \times L}$, where $d_{ij} = \mathbb{1}_{i=j} \sum_{j=1}^{L} \mathbb{1}_{\{a_{ij} > 0\}}, \forall i, j \in \mathcal{L}$, where $\mathbb{1}_{\{\cdot\}}$ denotes the indicator function. The connectivity of MSDs is determined by their coverage by one of the available MAPs. Figure 6.2 illustrates a typical scenario of two adjacent MAPs providing connectivity to the MSDs inside their influence region enabling network-wide connectivity. An MSD i is assumed to be connected to MAP j if it is closer to it than any other MAP, i.e. $\| y_i - q_j \| < \| y_i - q_k \|, k \in \mathcal{L} \backslash j$, and the MAP has sufficient capacity to serve the MSD. The quality of service or equivalently the utility of the communication link between an MSD i located at y_i and an MAP j located at q_j is represented by:

$$\Phi(i, j) = \phi(\| y_i - q_j \|), \tag{6.4}$$

where $\phi : \mathbb{R}^+ \rightarrow \mathbb{R}^+$ is continuously differentiable and decreasing function. The total number of MSDs connected to the MAPs is denoted by the vector $\mathbf{N}_u = [N_u^1, N_u^2, \ldots, N_u^L]^T$, while the maximum serving capacity of each MAP is denoted by $N^{\max}$. The distance based user association is motivated by the distance dependent signal decay. Each MSD aspires to connect to its nearest in-rage MAP unless constrained by the capacity of the host.

6.3 Dynamic Coverage and Connectivity Mechanism

In this section, we describe the methodology used to develop the cognitive and resilient connectivity framework for remotely deployed IoT devices. The locations of the MSDs are assumed to be constantly changing and are considered to be beyond the control of the MAPs. Our objective is to autonomously configure the MAPs in a distributed manner to provide coverage to the MSDs as well as

---

4 Note that this is not a norm but a mapping from a vector space to a scalar.

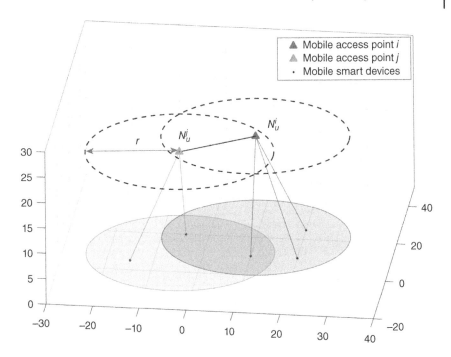

Figure 6.2 An example of two connected MAPs serving the underlying MSDs. The communication range of each MAP is depicted by the dotted lines while the area of influence is represented by the shaded circles.

keeping them connected to other MAPs. The cognitive connectivity resilience framework can be summarized by the cognition loop illustrated in Figure 6.3. The individual blocks of the cognitive framework are elaborated in the subsequent subsections.

6.3.1 MAP–MSD Matching

At each iteration of the cognitive connectivity framework, there needs to be an association between the MAPs and the MSDs. Since the wireless channel experiences distance dependent path loss, it is reasonable to assume a utility based on the Euclidean distance between the MSD and the MAP as given by (6.4). The preference of MSD i to connect to a MAP can be described as follows:

$$v(i) = \max_{j \in \mathcal{L} : \|y_i - q_j\| < r} \{\Phi(i,j)\}. \tag{6.5}$$

Notice, that only the MSDs that are under the influence of the MAPs are matched and the uncovered MSDs remain un-matched. The optimal assignment results in the matrix ε, where $\varepsilon_{ij} = \mathbb{1}_{\{j=v(i)\}}, \forall i \in \mathcal{M}, j \in \mathcal{L}$. As

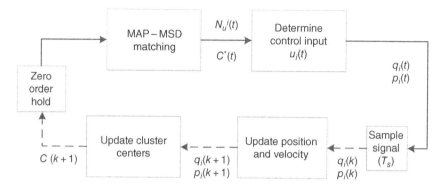

Figure 6.3 Feedback cognitive loop for the proposed resilient connectivity framework. Each MAP computes its control input based on the information about the location of MSDs and the number of connected users. The states of the MAPs are then updated in discrete time before repeating the observation cycle.

a result, the number of MSDs matched to each MAP can be evaluated as $N_u^i = \sum_{j=1}^{M} \epsilon_{ij}, \forall i \in \mathcal{L}$.

6.3.2 MAP Dynamics and Objective

We employ the kinematic model for the MAPs, whose dynamics can be written as follows:

$$\dot{q}_i = p_i,$$
$$\dot{p}_i = u_i, \qquad\qquad (6.6)$$

where $q_i, p_i, u_i \in \mathbb{R}^2, i \in \mathcal{L}$ represent the displacement, velocity, and acceleration of the devices, respectively. It can be represented using the state space representation $\dot{\mathbf{x}}_i = \tilde{A}_c \mathbf{x}_i + \tilde{B}_c \mathbf{u}_i$, where the vectors and matrices are defined as follows:

$$\mathbf{x}_i = \begin{bmatrix} q_i \\ p_i \end{bmatrix}, \quad \mathbf{u}_i = \begin{bmatrix} 0 \\ u_i \end{bmatrix}, \quad \tilde{A}_c = \begin{bmatrix} 0 & 1 \\ 0 & 0 \end{bmatrix}, \quad \tilde{B}_c = \begin{bmatrix} 0 \\ 1 \end{bmatrix}.$$

Note that the continuous time dynamical system is completely controllable. For practical implementation, the displacement and velocity vectors are discretized with a sampling interval of T_s. The equivalent discrete time system can be written as follows:

$$\mathbf{x}(k+1) = \tilde{A}_d \mathbf{x}(k) + \tilde{B}_d \mathbf{u}(k), \qquad\qquad (6.7)$$

where the matrices $\tilde{A}_d$ and $\tilde{B}_d$ govern the discrete time dynamics of the system.

The goal is to design a control input u_i for each MAP, which eventually leads to a desired configuration. To achieve this, we build upon the framework developed in [105] for distributed multi-agent systems and provide modifications, which lead to the desired behavior of MAPs in the context of D2D wireless networks constrained by the underlying MSDs. To enhance spatial coverage, the MAPs should have less coverage overlap and should be spread out while remaining connected to other MAPs. Therefore, we define a minimum distance $0 \leq d < r$ such that two MAPs should not be closer to each other than d unless they are forced to be closely located to serve a higher density of underlying MSDs.

6.3.3 Controller Design

We propose a control input $\mathbf{u} = [u_1, u_2, \ldots, u_L]^T$ for each of the MAPs in the following form:

$$u_i = f_i(\mathbf{q}, \mathbf{A}, \mathbf{N}_u) + g_i(\mathbf{p}, \mathbf{A}) + h_i(\mathbf{q}, \mathbf{p}), \tag{6.8}$$

where $f_i(\mathbf{q}, \mathbf{A}, \mathbf{N}_u)$ defines the gradient based term based on the attractive and repulsive forces between the MAPs, $g_i(\mathbf{p}, \mathbf{A})$ is the velocity matching term that forces neighboring MAPs to move with the same speed, and $h_i(\mathbf{q}, \mathbf{p})$ is the term defining the individual goals of each of the MAPs. Note that $\mathbf{q}$, $\mathbf{p}$, and $\mathbf{A}$ are functions of time. Such controllers holistically model the control of autonomous agents and are hence widely used in literature for a variety of robot swarming applications. Each of the terms in (6.8) is elaborated as follows:

6.3.3.1 Attractive and Repulsive Function

As highlighted earlier, the MAPs tend to maintain a minimum distance d with other MAPs unless the serving capacity is exceeded. Therefore, a repulsive force is required from MAP j to MAP i if the distance between them is less than d and an attractive force is needed from MAP i to MAP j if MAP j exceeds the capacity of serving MSDs. In effect, MAP i tends to share the load of MAP j if it exceeds capacity in order to provide coverage to all the MSDs. Therefore, the ith element of $f_i(\mathbf{q}, \mathbf{A}, \mathbf{N}_u)$ can be defined as follows:

$$f_i(\mathbf{q}, \mathbf{A}, \mathbf{N}_u) = \sum_{j \in \mathcal{N}_i} \left[\Psi(\| q_j - q_i \|_\sigma) \right.$$
$$\left. + a \left(1 - \alpha_{\{0,1\}} \left(\frac{\| (N_u^j - N^{\max})^+ \|_\sigma}{\| N^{\max} \|_\sigma} \right) \right) \right] \mathbf{v}_{ij}, i \in \mathcal{L}, \tag{6.9}$$

where $\mathbf{v}_{ij} = \nabla \| q_j - q_i \|_\sigma$ represents the vector in the direction going from the MAP at location q_i to the MAP at location q_j. The function $\Psi(z)$ is provided as follows:

$$\Psi(z) = \alpha_{\{\gamma,1\}} \left(\frac{z}{\| r \|_\sigma} \right) \phi(z - \| d \|_\sigma), \tag{6.10}$$

where $\phi(z) = \frac{1}{2}[(a+b)\frac{(z+c)}{\sqrt{1+(z+c)^2}} + (a-b)]$ is an un-even sigmoid function with $c = |a-b|/\sqrt{4ab}$ such that $\phi(z) \in (-a, a)$. Notice that the function $\Psi(z)$ is a product of two functions that results in the property that $\Psi(z) \leq 0$ if $z < \|d\|_\sigma$ and $\Psi(z) = 0$ otherwise, i.e. it provides a repelling force if MAP i and MAP j are closer than d and is neutral if they are farther than d. Therefore, the first term in the multiplier of the gradient in (6.9) ensures that the distance between neighboring MAPs is at least d. The second term is related to the attraction between MAPs if the MSDs aspiring to connect to them are beyond their capacity, i.e. $N^{\max}$. The force depends on the number of unserved users $(N_u^j - N^{\max})^+$ normalized by the maximum capacity and is accomplished using the function $a(1 - \alpha_{[0,1]}(\cdot))$ that maps from 0 to a, which is nonzero for strictly positive arguments.

6.3.3.2 Velocity Consensus Function

The velocity consensus function enables a matching between the velocities of neighboring MAPs and is expressed as follows:

$$g_i(\mathbf{p}, \mathbf{A}) = \sum_{j \in \mathcal{N}_i} a_{ij}(p_j - p_i), \quad i \in \mathcal{L}. \tag{6.11}$$

The function implies that MAP i tends to align its velocity to its neighbors weighted by the strength of the links. It generates a damping force in the movement of each of the MAP preventing any erratic behavior of any of them resulting in an unwanted collision. Furthermore, in the case where MSDs are moving in a coordinated fashion, the velocity consensus function prevents potential disconnections among the MAPs due to the otherwise relative velocity.

6.3.4 Individual Goal Function

The individual goal function $h(\mathbf{q}, \mathbf{p})$ is defined as follows:

$$h_i(\mathbf{q}, \mathbf{p}) = c_1(q_i^r - q_i) + c_2(p_i^r - p_i), \quad i \in \mathcal{L}, \tag{6.12}$$

where q_i^r and p_i^r are the reference position and velocity of MAP i, and c_1 and c_2 are positive constants denoting the relative aggressiveness to achieve the goal. It serves as a navigational feedback term that determines the eventual state that the MAPs aspire to achieve. If the goal of each MAP is precisely determined, the network can be made to reach the desired configuration. Assuming that each MAP is greedy to serve MSDs, a natural goal is to reach the centroid of the MSDs to allow a maximum number of MSDs to connect to it. Since the MSDs may be arbitrarily clustered, it is more efficient for the MAPs to move toward the centroid that is nearest to them. Hence the individual reference signals are selected as follows:

$$q_i^r = C_i^*, \quad \forall i \in \mathcal{L},$$
$$p_i^r = 0, \quad \forall i \in \mathcal{L}, \tag{6.13}$$

where C_i^* denotes the coordinates of the cluster center nearest to MSD i and is further elaborated in the sequel. A reference velocity of 0 implies that each MAP wants to eventually come to rest. The proposed cognitive loop propagates as follows: Given the spatial locations of the MSDs, a matching is made between the MAPs and the MSDs based on the distances and the maximum capacity of the MAPs. Based on the MAP–MSD association and the cluster centers (determined by the spatial location of MSDs), a control input is computed by each MAP independently and the system states are updated according to the dynamics provided by (6.6). After the locations and velocities are updated, new cluster centers of the MSDs are computed as their spatial locations might have changed due to mobility. Upon convergence, the control input becomes nearly zero and there is no further change in the configuration provided the MSDs do not change their positions.

6.3.5   Cluster Centers

In order to determine the destination of each MAP, we need information about the locations of the MSDs. Since the MSDs can move arbitrarily, they may not have a definitive spatial distribution. Therefore, it is reasonable to cluster the MSDs and use their centers as a potential destination for nearby MAPs. At each step, the objective is to find the following:

$$\underset{S}{\arg\min} \sum_{i=1}^{K} \sum_{y \in S_i} \| y - \overline{C}_i \|^2, \qquad (6.14)$$

The resulting optimal cluster centers of S are denoted by $C = [C_1, C_2, \ldots, C_K]^T$, $C_i \in \mathbb{R}^2$, $\forall i = 1, \ldots, K$. The solution to this problem can be obtained efficiently using Lloyd's algorithm [80]. In our proposed model, the only centralized information needed is the coordinates of the cluster centers of the MSDs. It can either be obtained using a centralized entity such as the satellite or it can be locally estimated based on individual observations. However, if local observations are used, then a distributed consensus needs to be made regarding the final goal state of each MAP. Due to the additional complexity in distributed consensus development, we have postponed it for future work. However, interested readers are directed to [96] and [95] for more details.

The flow of the cognitive connectivity algorithm is formally summarized in Algorithm 6.1. At epoch, each MAP obtains information about the centroid of the users known as the cluster centers and determines the coordinates of its nearest centroid denoted by C_i^*. Each of the MAPs interacts with the users in their area of influence and agree on serving the selected MSDs. The number of users supported by each MAP at time t is denoted by $N_u^i(t)$. Once the MAPs know their individual information, they exchange the information about the position, velocity, and number of connected users with their immediate

---

Algorithm 6.1 Resilient Connectivity Algorithm

Require: Initial position and velocity of each MAP $q(0)$ and $p(0)$; Initial K centroids of the MSDs $C(0)$; Time step Δ.

1: **repeat**
2:    Determine the coordinates of the nearest cluster center (goal position) for each MAP.
3:    Obtain a matching between each MAP and the MSDs in its area of influence using (6.4) and (6.5).
4:    Determine the number of connected users of each MAP, $N_u^i(t)$.
5:    Each MAP shares the position, velocity, and number of connected MSDs information with its immediate communication neighbors.
6:    Using $C(t)$ and $N_u^i(t)$, determine control input for each MAP $\mathbf{u}_i(t), i \in \mathcal{L}$ from (6.8).
7:    Update the position and velocity of each MAP using the discretized state space model $\mathbf{x}(k+1) = \tilde{A}\mathbf{x}(k) + \tilde{B}\mathbf{u}(k)$ with a time step of Δ.
8: **until** End of operation.

---

neighbors. Using the aggregated information, each MAP computes its control input $\mathbf{u}_i$. The position and velocity are then updated using the discrete time state space model in (6.7) with a time step of Δ. This process is repeatedly executed until the operation of MAPs is terminated. In the subsequent section, we describe the key metrics that are used to evaluate the performance of the proposed cognitive algorithm.

6.4 Performance Evaluation and Simulation Results

In order to evaluate the performance of the proposed cognitive connectivity framework, we use the following metrics to measure the connectivity of the network:

1. **Proportion of MSDs Covered**: It measures the percentage of total MSDs that are associated and served by one of the MAPs. This metric helps in determining the general accessibility of the MSDs to the MAPs. However, it does not reflect the true connectivity of the MSDs as the MAPs might not be well connected.

2. **Probability of Information Penetration in MAPs**: Assuming that the connected MSDs can communicate perfectly with the MAPs, the overall performance of the system depends on the effectiveness of communication between the MAPs using D2D links. One way to study the dynamic information propagation and penetration in D2D networks is based on mathematical epidemiology (see [39]). Since the network in this chapter is

finite, we make use of the N-intertwined mean field epidemic model [141] to characterize the spread of information. The steady state probability of MAP i being informed by a message, denoted by $v_{i\infty}$, propagated in the D2D network at an effective spreading rate of τ is bounded as follows (see Theorem 1 in [141]):

$$0 \leq v_{i\infty} \leq 1 - \frac{1}{1 + \tau d_{ii}}. \tag{6.15}$$

It measures the probability of a device being informed about a piece of information that initiates from any device in the network at random. In other words, a 70% probability of information dissemination reflects that if a piece of information is generated at random by any of the nodes, there is a 70% probability on average than any other node in the network will receive it assuming perfect success in the transmissions.

3. **Reachability of MAPs**: While it is important to determine the spreading of information over the D2D network, it is also crucial to know whether the D2D network is connected or not. It can be effectively determined using the algebraic connectivity measure from graph theory, also referred to as the *Fiedler value*, i.e. $\lambda_2(\mathbf{L})$, where $\lambda_2(\cdot)$ denotes the second-smallest eigenvalue and $\mathbf{L}$ is the Laplacian matrix of the graph defined by the adjacency matrix $\mathbf{A}$. A nonzero Fiedler value indicates that each MAP in the network is reachable from any of the other MAPs. The Laplacian matrix is defined as follows:

$$\mathbf{L} = \mathbf{D} - \mathbf{A}. \tag{6.16}$$

These metrics complement each other in understanding the connectivity of the network. The resilience, on the other hand, is measured in terms of the percentage of performance recovery after an event of failure has occurred. In Section 6.4.1, we provide the simulations results and describe the main observations in comparison with existing approaches.

6.4.1 Results and Discussion

6.4.1.1 Simulation Parameters
In this section, we first describe the simulation settings before providing results on the performance of our proposed cognitive connectivity framework. Note that the selection of simulation parameters is made for illustrative purposes. A bi-layer communication network is considered, comprising of MSDs such as IoT devices and MAPs such as UAVs. A set of $M = 2000$ MSDs are distributed in $\mathbb{R}^2$ according to a 2-D Gaussian mixture model with equal mixing proportions. The mean vectors are selected as $\mu_1 = [50, 20]^T, \mu_2 = [0, -50]^T$, and $\mu_3 = [-40, 40]^T$, and the covariance matrices are selected as follows:

$$\Sigma_1 = \begin{bmatrix} 200 & 0 \\ 0 & 100 \end{bmatrix}, \quad \Sigma_2 = \begin{bmatrix} 500 & 0 \\ 0 & 200 \end{bmatrix}, \quad \Sigma_3 = \begin{bmatrix} 150 & 0 \\ 0 & 300 \end{bmatrix}. \tag{6.17}$$

The mobility of the MSDs is modeled by a scaled uniform random noise at each time step, i.e. $y_i(k + 1) = y_i(k) + s\xi$, where $\xi \sim \text{Uniform}([-1, 1] \times [-1, 1])$, where $s = 0.2$ represents the scale. The $L = 80$ MAPs are initially distributed uniformly in the plane perpendicular to the vector $[0, 0, h]^T$ where the altitude h is selected as 20 m. The initial velocity vectors of the MAPs are selected uniformly at random from the box $[-1, -2]^2$. It is assumed that the utility function used to define the quality of the communication link between the MSDs and the MAPs is $\Phi(i, j) = \kappa \parallel y_i - q_j \parallel^{-\eta}$, where $\kappa \in \mathbb{R}^+$ is the transmission power and $\eta \in \mathbb{R}^+$ represents the path-loss exponent. Furthermore, we assume that $\kappa = 1$ and $\eta = 4$. A list of all the remaining parameter values used during the simulations is provided in Table 6.1. We run the simulations with a step size of $\Delta = T_s$ up to $t = 25$ seconds.

For the purpose of simulation, the strength of communication links between the MAPs and their neighbors is characterized by the cutoff function in (3) with thresholds $z_1 = \gamma = 0.2$ and $z_0 = 1$. In other words, we have abstracted the wireless propagation details by using the cutoff function $\alpha_{\{\gamma,1\}}$ which is parameterized by γ. Hence, each MAP only requires the knowledge of the distance to its neighboring MAPs that are inside its communication range and the threshold γ to evaluate its control strategy using (8). The particular choice of γ implies that two MAPs are considered to reliably communicate if the normalized distance between them is within 20% of the maximum normalized communication range. On the other hand, two MAPs are considered completely unable to communicate if the normalized distance is greater than the maximum normalized communication range. However, for practical implementation, the threshold γ needs to be accurately determined based on the radio propagation model used and the communication reliability requirements.

In Figure 6.4, we provide an example run of the proposed cognitive connectivity framework. We show a top view of the network for the sake of clear presentation. Figure 6.4a shows the initial configuration at $t = 0$, when the

Table 6.1 Simulation parameters.

Parameter	Value	Parameter	Value
M	2000	a	5
L	80	b	5
r	24	c_1	0.2
d	20	c_2	0.1
ϵ	0.1	s	0.2
$N^{\max}$	80	τ	1
h	20	T_s	0.01
k	3	γ	0.2

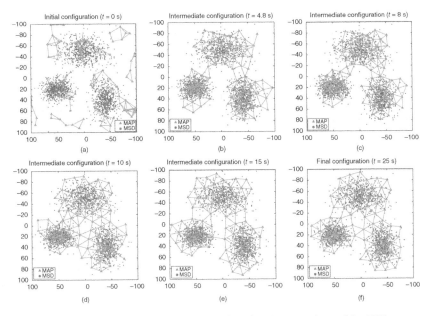

Figure 6.4 Example run of the cognitive algorithm showing snapshots of the MAP configuration at the initial stage, intermediate stages, and final stage after convergence.

MAPs are randomly deployed over an underlying population of MSDs. The MSDs keep moving with time in the goal directions as observed by the different MSD locations in Figure 6.4b–e. As the proposed cognitive connectivity framework evolves, the MAPs tend to move toward the closest group of MSDs as shown in Figure 6.4b. Finally, when the framework converges, the MAPs develop a desirable connected formation hovering over the MSDs as shown by Figure 6.4f. It should be noted that the MAPs are located closer to each other in areas where MSDs are densely deployed, such as around cluster centers. In areas where the MSDs are sparsely located, the MAPs develop a regular formation.

6.4.1.2 Resilience

In this section, we investigate the impact of a random MAP failure event on the connectivity of the network and evaluate the response of the proposed cognitive framework in such a situation. Figure 6.5 shows an induced random failure of 20% of the MAPs, which results in loss of coverage to the MSDs as well as reduction in the connectivity of the MAPs. The proposed framework immediately starts responding to the coverage gap created by the MAP failure and tends to reconfigure itself as shown by the intermediate snapshot in Figure 6.5b. Eventually, the framework converges leading to a coverage maximizing configuration while maintaining connectivity of the MAPs as shown in Figure 6.5c.

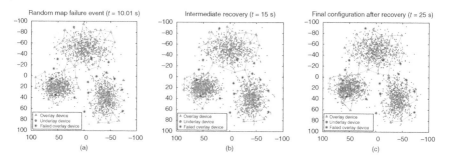

Figure 6.5 A random MAP failure event is induced at $t = 10$ seconds, making 20% of the MAPs unavailable. The cognitive framework adaptively re-configures itself to improve network connectivity as shown in the snapshots.

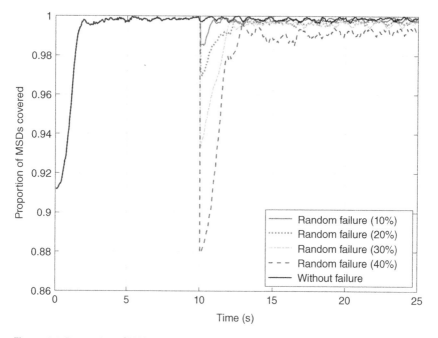

Figure 6.6 Proportion of MSDs covered by the MAPs.

To test the resilience of the proposed framework under different levels of MAP failure events, we simulate varying severity of device failure events from 10% to 40% failed MAPs in the overlay network. Figure 6.6 illustrates the proportion of MSDs that are covered by the MAPs. Without any failure, it is observed that the proposed framework successively improves the coverage until almost all the MSDs are covered. Notice that the coverage fluctuations occur due to the continuously mobile MSDs. Once the device failure event occurs at around

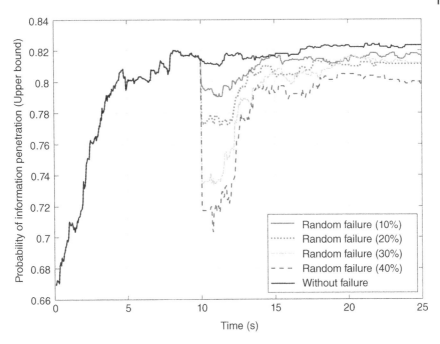

Figure 6.7 Probability of information penetration in the D2D enabled MAP network.

$t = 10$ seconds, the framework responds and is able to quickly restore maximum coverage except in the case of 40% failure, in which the coverage is not fully restored. In Figure 6.7, we plot the probability of information penetration in the MAP network. It can be observed that the framework is able to recover up to 97% of the original value. However, it is important to note that the probability of information dissemination is an upper bound and does not provide information about the reachability of the MAPs. In this situation, Figure 6.8, which shows the algebraic connectivity of the MAPs, proves to be extremely useful. It is observed that when the failure proportion is 10%, 20%, or 30%, the reachability can be restored by the proposed framework, as indicated by the nonzero algebraic connectivity. However, in the event of 40% failure, the algebraic connectivity remains zero even after reconfiguration, which implies that the MAP network is no more connected. However, since the probability of information penetration is still high, it implies that the MAPs have also clustered around the MSD clusters thus providing effective intra-cluster connectivity. Note that there is a spike in the MAP reachability at $t = 4.8$ seconds. This is because the dynamic MAPs momentarily become connected at $t = 4.8$ seconds as shown by Figure 6.5b before being disconnected by a single link.

Furthermore, the framework is also resilient to many other variations in the network. For instance, in the case of crowd mobility, the proposed cognitive

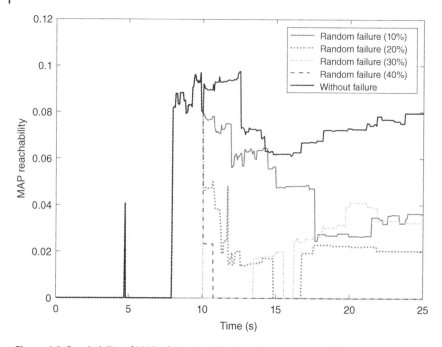

Figure 6.8 Reachability of MAPs determined by the algebraic connectivity.

framework can readily adapt to the changing positions of the users without additional complexity. However, existing approaches in literature cannot adapt to such changes in real-time. This is because they require global network information along with the need to solve the NP-hard optimization problem repeatedly. Similarly, the proposed framework is also resilient to a wide variety of cyber–physical attacks as it uses a cognitive feedback loop to constantly update the states of each MAP independently of the others, which makes it less vulnerable to cyber threats.

6.4.1.3 Comparison

In this section, we compare the performance of our proposed algorithm with existing frameworks. We use the two widely used approaches in literature, i.e. the p-median approach and the circle packing approach as baseline algorithms. Since the compared solution algorithms provide centralized solutions, we ignore the effect of mobility of the MSDs in this section for fair comparison. Before measuring the performance in terms of the metrics, we illustrate an example MAP configuration obtained using each of the approaches in Figure 6.9. Figure 6.9a shows the optimal configuration of MAPs obtained using the circle packing solution. It can be observed that the MAPs are tightly packed inside a circular region covering the MSDs. Figure 6.9b shows the

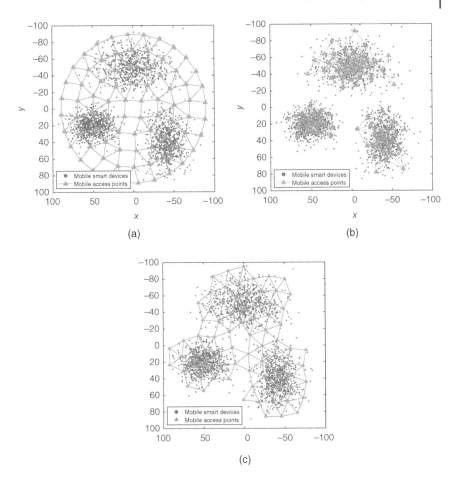

Figure 6.9 Top view of final configurations of MAPs in comparison with the proposed approach. (a) Circle packing approach, (b) p-median approach, and (c) proposed dynamic approach.

optimal configuration achieved using the p-median solution. As we would expect, all the MAPs are concentrated around the MSDs and since the MSDs are distributed in spatial clusters, the MAPs are unable to maintain connectivity between them. Finally, Figure 6.9c shows the MAP configuration achieved after convergence of our proposed cognitive framework. Note that our proposed framework leads to a configuration that covers all the MSDs and also maintains connectivity between the MAPs.

Next, we compare the performance of the proposed algorithm with the baseline in terms of the metrics set for performance evaluation. Figure 6.10 shows the proportion of MSDs covered by the MAPs for each of the algorithms. It can

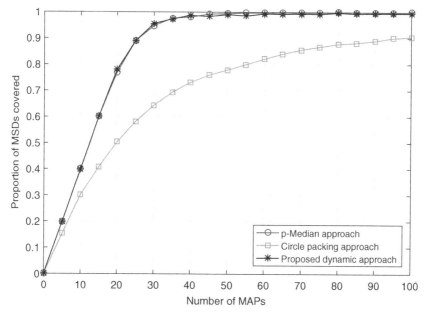

Figure 6.10 Comparison of coverage of MSDs with baseline algorithms.

be observed that the proposed dynamic algorithm and the p-median approach results lead to the complete coverage of the MSDs with significantly less number of MAPs (~20) as compared with the circle packing approach (>100). The primary reason for this is the restriction of MAPs to fill a circle as packing circles inside an arbitrary region is NP-hard. Moreover, since the MAPs cannot be positioned arbitrarily close to each other, so there is loss of coverages when the number of users under the influence of a MAP exceeds its capacity. In Figure 6.11, we plot the reachability or algebraic connectivity of the MAPs using the three algorithms. It can be observed that the p-median approach does not lead to a connected network even with very high number of MAPs. On the other hand, both the proposed algorithm and the circle packing approach leads to a connected configuration, i.e. the Fiedler value is non-zero, if the number of MAPs is sufficiently high (>40). Note that the magnitude of the Fiedler value reflects the tendency of the network to become disconnected. Since the circle packing approach leads to a well-connected configuration, it is harder to make the network disconnected, which is why it has a higher Fiedler value. Finally, Figure 6.12 shows the average probability of information dissemination in the MAP network. It can be observed that the proposed cognitive approach results in a high average information penetration probability with a small number of MAPs (~ 20). On the other hand, the other approach requires significantly

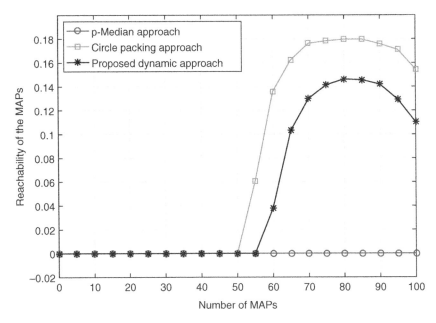

Figure 6.11 Comparison of reachability of MAPs with baseline algorithms.

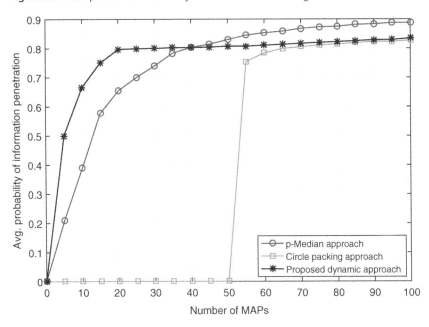

Figure 6.12 Comparison of probability of information dissemination in the MAPs with baseline algorithms.

higher number of MAPs to achieve a similar level of average information dissemination. In light of the earlier comparisons, it is clear that the proposed cognitive framework significantly outperforms other frameworks available in literature. It is also pertinent to mention that the baseline algorithms are not dynamic and hence would perform even worse under the mobility of MSDs.

6.5 Summary and Conclusion

In this chapter, a cognitive connectivity framework is presented that reconfigures itself autonomically in a distributed manner to interconnect spatially dispersed smart devices thus enabling the IoT in remote environments. Resilience of connectivity has been investigated in response to the mobility of the underlay network as well as random device failures in the overlay network. It is shown that if a sufficient number of overlay devices are available, then the developed distributed framework leads to high network connectivity which is resilient to mobility and device failures. However, if sufficient overlay devices are not deployed, the framework tends to provide connectivity locally to the devices in each cluster of the underlay network. A comparison of the proposed approach with existing approaches for placement of BSs reveals significant superiority in terms of the number of BSs required to achieve coverage and the overall connectivity of the devices.

This work provides a useful platform for the development of more sophisticated and efficient algorithms to achieve a variety of objectives in aerial communications using UAVs. Future directions in this work can investigate on ways to make the framework completely distributed. The local observations of MAPs can be used to form a consensus about the locations of the MSDs. Another possible direction to this line of research is to allow MAPs to operate in multiple modes to enable connectivity between a diverse pattern of locations of the MSDs.

Part IV

Secure Network Design Mechanisms

7

Wireless IoT Network Design in Adversarial Environments

7.1 Adversarial Network Scenarios

The interconnection of combat equipment and other battlefield resources for coordinated automated decisions is referred to as the Internet of battlefield things (IoBT). IoBT networks are significantly different from traditional Internet of things (IoT) networks due to battlefield specific challenges such as the absence of communication infrastructure, heterogeneity of devices, and susceptibility to cyber–physical attacks. The combat efficiency and coordinated decision-making in war scenarios depends highly on real-time data collection, which in turn relies on the connectivity of the network and information dissemination in the presence of adversaries. This work aims to build the theoretical foundations of designing secure and reconfigurable IoBT networks. Leveraging the theories of stochastic geometry (SG) and mathematical epidemiology, we provide an integrated framework to quantify the information dissemination among heterogeneous network devices. Consequently, a tractable optimization problem is formulated that can assist commanders in cost effectively planning the network and reconfiguring it according to the changing mission requirements.

7.2 Modeling Device Capabilities and Network Heterogeneity

In this section, we first describe the geometry of the IoBT network and propose a bi-layer abstraction model using tools from SG. Then, we provide a dynamic model to characterize information dissemination in the heterogeneous IoBT network based on mathematical epidemiology.

Resource Management for On-Demand Mission-Critical Internet of Things Applications, First Edition.
Junaid Farooq and Quanyan Zhu.
© 2021 John Wiley & Sons, Inc. Published 2021 by John Wiley & Sons, Inc.

7.2.1 Network Geometry

Consider uniformly deployed heterogeneous battlefield things in $\mathbb{R}^2$ that are abstracted as a Poisson point process (PPP)[1] [133] with intensity λ devices/km^2, referred to as $\Phi = \{\mathcal{X}_i, \mathcal{T}_i\}_{i \geq 1}$, where $\mathcal{X}_i$ and $\mathcal{T}_i$ represent the location and type of the ith device, respectively. We assume that the network is composed of two types of devices, i.e. $\mathcal{T}_i \in \{1, 2\}, \forall i \geq 1$. The first type of devices, i.e. $\mathcal{T}_i = 1$, referred to as *type*-I devices, are equipped with two radio interfaces. The second type of devices, i.e. $\mathcal{T}_i = 2$, referred to as *type*-II devices, have only one radio interface that is compatible with type-I devices. In the first network layer, only type-I devices can communicate with other type-I devices, while in the second layer, both type-I and type-II devices can communicate with each other due to the availability of common radios. Assuming that each device can be of type-I with probability p, i.e. $\mathbb{P}(\mathcal{T}_i = 1) = p, \forall i \geq 1$, then the set of active devices in the first network layer can be represented by a PPP $\Phi_1 = \{\mathcal{X}_i \in \Phi : \mathcal{T}_i = 1\}$ with intensity $\lambda_1 = p\lambda$, which is obtained by an independent thinning of the original point process Φ. On the other hand, since all the devices in the second network layer can communicate with each other, the active devices can be represented by a PPP $\Phi_2 = \Phi$, with intensity $\lambda_2 = \lambda$. Note that the devices can move independently and we assume that the placement of devices as a result of mobility remains uncorrelated, i.e. can be represented by a new realization of the original PPP, thus resulting in a quasi-static network.

This type of network configuration is particularly well-suited to IoBT networks, where type-I devices may correspond to ground stations or armored vehicles equipped with multiple types of radios, while type-II devices may correspond to soldiers equipped with single radio smart mobile devices. The equipped radios on the devices are characterized by their transmission power or equivalently, the communication range r_m in meters, $m \in \{1, 2\}$. Type-I devices have two radios with transmission ranges r_1 and r_2, respectively, while type-II devices have one radio with transmission range r_2. The communication range of the radios is tunable in the interval $[r_m^{\min}, r_m^{\max}]$, where $r_m^{\min} \geq 0$ and $r_m^{\max} \geq r_m^{\min}, \forall m \in \{1, 2\}$ and we assume that $r_2 \leq r_1$.

Due to the absence of traditional communication infrastructure such as base stations, the devices are only able to communicate using device-to-device (D2D) communications, i.e. devices $x, y \in \Phi_m$ are connected to each other in network layer m if $\|x - y\| \leq r_m$, $m \in \{1, 2\}$, where $\| \cdot \|$ represents the Euclidean distance. Similarly, devices $x \in \Phi_1$ and $y \in \Phi \backslash \Phi_1$ can communicate with each other only if $\|x - y\| \leq r_2$. Hence, the communication links between devices in each layer can be modeled using a random geometric graph (RGG) [28] with a connection radius of $r_m, m \in \{1, 2\}$. Each of the layers of the

---

1 While the devices can be placed more strategically according to their characteristics and utility, however, due to potential mobility and difficulty of tracking network topology, these optimal locations may not be known or fixed, which justifies the PPP assumption.

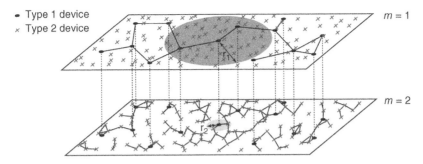

Figure 7.1 Heterogeneous IoBT network decomposed into virtual connectivity layers. The dark gray region illustrates the communication reach of type-I devices in layer 1, while the light gray region illustrates the communication reach of type-I and type-II devices in layer 2.

multiplex network has a different connectivity that depends on the respective device densities and the communication ranges. An illustrative representation of the network model is provided in Figure 7.1 that shows the connectivity between type-I devices in the first layer and the connectivity between all devices in the second layer. In the subsequent subsection, we analyze the connectivity of the devices in each layer, referred to as *intra-layer* connectivity and the connectivity of the overall network, referred to as *network-wide* connectivity.

7.2.2 Network Connectivity

In this section, we describe the connectivity between the heterogeneous devices in an IoBT network. The connectivity of devices can be classified into *intra-layer* and *network-wide* connectivity, which are explained as follows:

7.2.2.1 Intra-layer Connectivity

Within a particular network layer m, the active[2] devices can communicate with each other if they are within a distance of r_m of each other. The set of communication neighbors of a typical device $x \in \Phi_m$ in layer m can be expressed as $\mathcal{N}_m(x) = \{y \in \Phi_m : \|x - y\| \leq r_m\}$. The connectivity of the RGG formed by devices in layer m is characterized by the degree of the devices, denoted by K_m, which is defined as the number of neighbors of each device, i.e. $K_m = |\mathcal{N}_m(x)|$, where $|\cdot|$ represents the set cardinality. Since the network is spatially distributed as a PPP, the degree of each device in the RGG is a Poisson random variable [58]. Therefore, the resulting intra-layer degree distribution of a typical device can be expressed by the following lemma.

---

2 Type-I devices are active in both network layers, while type-II devices are only active in the second network layer.

Lemma 7.1 *The intra-layer degree of a typical device in each network layer is distributed as follows:*

$$\mathbb{P}(K_1 = k) = \begin{cases} (1-p) + pe^{-\lambda_1 \pi r_1^2}, & \text{if } k = 0, \\ pe^{-\lambda_1 \pi r_1^2} \dfrac{(\lambda_1 \pi r_1^2)^k}{k!}, & \text{if } k > 0, \end{cases} \tag{7.1}$$

$$\mathbb{P}(K_2 = l) = e^{-\lambda_2 \pi r_2^2} \frac{(\lambda_2 \pi r_2^2)^l}{l!}, \quad l \geq 0, \tag{7.2}$$

for sufficiently large $\lambda_1, \lambda_2, r_1,$ and r_2. The average degree of a typical device in the two network layers can be expressed as follows:

$$\mathbb{E}[K_1] = p\lambda_1 \pi r_1^2, \tag{7.3}$$

$$\mathbb{E}[K_2] = \lambda_2 \pi r_2^2. \tag{7.4}$$

Proof: To evaluate the degree distribution of the devices in the first network layer, we proceed as follows:

$$\mathbb{P}(K_1 = k) = \sum_{j=1}^{2} \mathbb{P}(K_1 = k | \mathcal{T}_i = j)\mathbb{P}(\mathcal{T}_i = j). \tag{7.5}$$

Since type-II devices are not active in the first network layer, we know that $\mathbb{P}(K_1 = 0 | \mathcal{T}_i = 2) = 1$ and $\mathbb{P}(K_1 > 0 | \mathcal{T}_i = 2) = 0$. However, type-I devices can communicate with other type-I devices in the first network layer. Since the active devices in the first layer are also represented as a PPP, the degree of the RGG formed among the active devices is Poisson distributed [58], i.e. $\mathbb{P}(K_1 = k | \mathcal{T}_i = 1) \sim \text{Poisson}(\lambda_1 \pi r_1^2)$. It follows that $\mathbb{P}(K_1 = 0 | \mathcal{T}_i = 1) = e^{-\lambda_1 \pi r_1^2}$ and $\mathbb{P}(K_1 > 0 | \mathcal{T}_i = 1) = \frac{e^{-\lambda_1 \pi r_1^2}(\lambda_1 \pi r_1^2)^k}{k!}, k > 0$. Finally, since $\mathbb{P}(\mathcal{T}_i = 1) = p$ and $\mathbb{P}(\mathcal{T}_i = 2) = 1 - p$, we can substitute the corresponding expressions for $K_1 = 0$ and $K_1 > 0$ in (7.5) to obtain the result in (7.1). Similarly, for the distribution of K_2, we know that $\mathbb{P}(K_2 = k) \sim \text{Poisson}(\lambda_2 \pi r_2^2)$ regardless of $\mathcal{T}_i$. This directly leads to the result in (7.2). Using these probability distributions, it is straightforward to derive the expected degree in both layers given by (7.3) and (7.4). $\square$

From Figure 7.1, it is clear that the average degree or equivalently the connectivity of devices in each layer depends on the density of the deployed devices as well as the communication range. The joint probability distribution of the connectivity of a typical device in each layer is denoted by $\mathbb{P}(K_1 = k, K_2 = l)$.

7.2.2.2 Network-wide Connectivity

If the two network layers are collapsed together to form a single virtual network such that both layers reinforce the connectivity of the devices, then the connectivity is characterized in terms of the combined degree denoted by K_c. The combined-layer degree of a typical device $x \in \Phi$ is defined as $K_c = |\mathcal{N}_1(x)| + |\mathcal{N}_2(x)|$. Since the degree of the devices in each layer is Poisson distributed, the combined-layer degree follows a Poisson mixture distribution expressed by the following lemma.

Lemma 7.2 *The average combined-layer degree of a typical device can be expressed as follows:*

$$\mathbb{E}[K_c] = p\lambda_1 \pi r_1^2 + \lambda_2 \pi r_2^2. \tag{7.6}$$

Proof: The distribution of the combined degree can be evaluated as follows:

$$\mathbb{P}(K_c = k) = \sum_{j=1}^{2} \mathbb{P}(K_c = k | \mathcal{T}_i = j)\mathbb{P}(\mathcal{T}_i = j). \tag{7.7}$$

It is clear that $\mathbb{P}(K_c = k | \mathcal{T}_i = 2) \sim \text{Poisson}(\lambda_2 \pi r_2^2)$ as type-II devices are only connected in the second network layer. However, if a typical device is of type-I, then the degree needs to be more carefully evaluated. We have provided an illustration in Figure 7.2 to aid in characterizing the degree. The degree of a type-I device placed at the origin can be expressed as $K_c = 2N_{1A} + N_{2A} + N_{1B}$, where N_{1A} and N_{2A} represent the number of devices of type-I and type-II, respectively, in the circular region labelled **A**. N_{1B} represents the number of type-I devices in the hollow circular region labelled **B**. Note that type-I and type-II devices are distributed according to independent PPPs with intensity $p\lambda$ and $(1-p)\lambda$. It results in the fact that N_{1A},

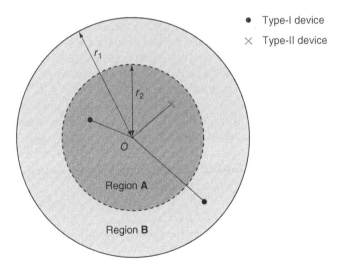

Figure 7.2 Combined degree of a typical device located at the origin.

N_{2A}, and N_{1B} are independent random variables following a Poisson($p\lambda\pi r_2^2$), Poisson($(1-p)\lambda\pi r_2^2$), and Poisson($p\lambda\pi(r_1^2 - r_2^2)$), respectively. The sum K_c is not composed of independent terms, however, due to N_{1A} that has a multiplicity of 2. Therefore, $\mathbb{P}(K_c = k | \mathcal{T}_i = 1) = \mathbb{P}(N_{1A} = \frac{k}{2}) * \mathbb{P}(N_{2A} = k) * \mathbb{P}(N_{1B} = k)$, where $*$ represents the convolution operator. Therefore, the probability distribution of the combined degree can be obtained by substituting the previously developed expressions into (7.7) and using the fact that $\mathbb{P}(\mathcal{T}_i = 1) = p$ and $\mathbb{P}(\mathcal{T}_i = 2) = 1 - p$. The average combined layer degree can be obtained as $\mathbb{E}(K_c) = p(2\mathbb{E}(N_{1A}) + \mathbb{E}(N_{2A}) + \mathbb{E}(N_{1B})) + (1-p)(\lambda\pi r_2^2) = p\lambda_1\pi r_1^2 + \lambda_2\pi r_2^2$. □

In the following section, we describe the information dissemination over such bi-layer networks that have the aforementioned connectivity profile.

7.3 Information Dissemination Under Attacks

Each type of device in the IoBT network generates data that needs to be propagated to other devices of the same type and/or different types of devices depending on the role of that device. There are certain pieces of information that needs to be shared among the same type of devices, e.g. soldiers need to communicate information with other soldiers and similarly commanding units might also share information amongst them. Henceforth, we refer to the information sharing in each network layer as *intra-layer information dissemination*. On the other hand, some information might be important for all network nodes such as network health monitoring data or network discovery beacons. This is henceforth referred to as *network-wide information dissemination*. We assume a time slotted system where the duration of each slot is τ s. During each time slot, the informed devices broadcast information to their neighbors at a rate of γ. Let $\mathcal{P}_s^{(i)} \in [0, 1]$ be the average probability that the transmitted information type $i \in \{1, 2, c\}$ by a typical device is successfully received by its neighbors, referred to as the *success probability*, and $\delta \in [0, 1]$ be the probability that the communication will be affected by cyber–physical attacks. Since the event of successful transmission due to interference from other devices and the event of a cyber–physical attack are independent, the effective probability of a successful transmission can be expressed as $(1 - \delta)\mathcal{P}_s^{(i)}$. Consequently, the information spreading rate between devices, denoted by $\alpha^{(i)}, i \in \{1, 2, c\}$, can be expressed as follows:

$$\alpha^{(i)} = \gamma(1 - \delta)\mathcal{P}_s^{(i)}. \tag{7.8}$$

We refer to δ as the *threat level* as it signifies the perceived risk in information transmission between devices. Without loss of generality,[3] we select the

---

3 There is no loss of generality since τ can be made arbitrarily small.

contact rate $\gamma = 1$, so effectively, $\alpha^{(i)}$ is the probability of successful information transmission between devices. In essence, $\alpha^{(i)}$ can be interpreted as the desired security level from the perceived threats to the communication network. The probability of successful transmission can be computed by setting a threshold on the received signal-to-noise-plus interference-ratio (SINR) at a typical device. Several techniques can be used from SG literature to accurately characterize the SINR based success probability depending on the medium access protocol used [11]. We use a generalized representation of the success probability $\mathcal{P}_s^{(i)} = g_i(p, \lambda, \mathbf{r})$ in terms of the densities and communication ranges of the devices, where the function $g_i(\cdot)$ is assumed to be monotone in its arguments. The parameter δ can capture a broad range of cyber-physical threats in IoBT networks. Different methods can be used to assess the threat level in battlefields due to jamming, physical attacks, and other adversarial actions based on historical data and/or statistical models of attack types, some of which are explored in existing works such as [92] and [117]. For instance, to model jamming attacks, the parameter δ can be based on the SINR, in which case the RGG becomes an interference graph [35]. To tackle physical network attacks such as targeted attacks, the parameter δ can be based on the density of device deployment, the connectivity of devices, or the type of devices. Furthermore, an integrated metric can also be developed that can simultaneously capture a multitude of threats. Over time, the adversarial attacks may compromise or negatively impact significant portion of the network connectivity. So there is a need for a resilient framework that can be reconfigured to recover from the lost connectivity by cyber–physical attacks.

Another important aspect of the information dissemination process is to account for the information annihilation at each time step. There are several reasons that a device may not broadcast the information that it has received from another device in the previous time slot such as limited buffer capacity and misclassifying information as unimportant. However, the most important factor is to ensure propagating the most recent information in the network. This dynamical information spreading process in an IoBT network can be formalized using the susceptible-infected-susceptible (SIS) model [109], which is well studied in mathematical epidemiology. The challenge is that information propagates over a topology in wireless communication networks, while the classical SIS model does not deal with topological constraints. Furthermore, we deal with simultaneous information dissemination in multiple network layers, which present additional challenges. In Section 7.3.1, we describe the dynamics of the information dissemination process.

7.3.1 Information Dynamics

In this section, we present the dynamics of information dissemination across the IoBT network. For ease of explanation, we first describe the dynamics of a single message being propagated in the network, followed by the dynamics of two messages simultaneously spreading in the network.

7.3.1.1 Single Message Propagation

If all the devices in the network disseminate the same message from one device to another in a broadcast manner during each time slot using all the available radio interfaces, then each device with degree k can be in either an uninformed state (U_k) or an informed state (I_k) depending on the success of information transmission. To model this behavior and explain the dynamics of information dissemination across the IoBT network, we exploit the SIS model [109] from mathematical epidemiology. The information dissemination is directly related to the degree of the devices in the network, which in turn depends on the physical network parameters. Since the network is random with potentially a large number of devices, we use the degree based mean-field approach, in which all devices are considered to be statistically equivalent in terms of the degree and the analysis is done on a typical device. Therefore, the information dissemination dynamics of the system can be written in terms of the degree of the devices as follows [90]:

$$\frac{dI_k^{(c)}(t)}{dt} = -\mu I_k^{(c)}(t) + \alpha^{(c)} k U_k^{(c)}(t) \Theta^{(c)}, \tag{7.9}$$

where $I_k^{(c)}(t)$ denotes the proportion of devices with degree k that are in state I_k, i.e. informed with network-wide information, at time t, $U_k^{(c)}(t) = 1 - I_k^{(c)}(t)$ denotes the proportion of degree k devices that are in state U_k. The superscript (c) refers to the combined-layer network signifying network-wide information dissemination. The first term in (7.9) explains the annihilation of information with time, i.e. the informed devices return to the uninformed sate at a rate of μ. This ensures that at each time step, only the most recent information is propagated in the network as a particular piece of information is discarded after being retransmitted multiple times depending on the annihilation rate. For equilibrium analysis, we can assume that $\mu = 1$ as the effect of the annihilation can be captured by the effective spreading rate. The second term accounts for the creation of informed devices due to the spreading. The rate of increase in the density of informed devices with degree k is directly proportional to the degree, the probability of successful transmission of information α, the proportion of uninformed devices with degree k, i.e. $U_k^{(c)}(t)$, and the average probability that a neighbor of a device with degree k is informed, denoted by $\Theta^{(c)}$. The probability $\Theta^{(c)}$ can be computed as $\sum_{k'} \mathbb{P}(K_c^{\text{neighbor}} = k' | K_c = k) I_{k'}^{(c)}(t)$, where $\mathbb{P}(k' | K_c = k)$ denotes the probability that a neighbor of a typical device with $K_c = k$ has a degree $K_c^{\text{neighbor}} = k'$. Since the network is PPP, the degrees are uncorrelated, i.e. $\mathbb{E}[K_c^{\text{neighbor}} K_c] = \mathbb{E}[K_c^{\text{neighbor}}]\mathbb{E}[K_c]$, we can effectively write $\mathbb{P}(K_c^{\text{neighbor}} = k' | K_c = k) I_{k'}^{(c)}(t) = \frac{k' \mathbb{P}(K_c^{\text{neighbor}} = k')}{\mathbb{E}[K_c]} I_{k'}^{(c)}(t)$. Hence, by appropriately renaming dummy variables, $\Theta^{(c)}$ can be expressed as follows [109]:

$$\Theta^{(c)} = \sum_{k \geq 0} \frac{k \mathbb{P}(K_c = k)}{\mathbb{E}[K_c]} I_k^{(c)}(t), \tag{7.10}$$

where $\mathbb{E}[K_c]$ is provided in Section 7.2.2. Note that another expression for the rate of change in the uninformed devices, i.e. $\frac{dU_k^{(c)}(t)}{dt}$, can also be written; however, it is not useful since $U_k^{(c)}(t)$ depends directly on $I_k^{(c)}(t)$ and thus (7.9) completely describes the dynamics of network-wide information dissemination.

7.3.1.2 Multiple Message Propagation

In general, there may be multiple messages or pieces of information spreading in the IoBT network at any particular time. However, for the bi-layer network model, we assume that there are two messages propagating in the network, i.e. one in each network layer. Therefore, each device with a degree k in the first layer and l in the second layer can be in one of the four possible dynamical states, i.e. uninformed of both messages ($UU_{k,l}$), informed of message 1 but uninformed of message 2 ($IU_{k,l}$), uninformed of message 1 but informed of message 2 ($UI_{k,l}$), and informed of both messages ($II_{k,l}$). These state variables denote the proportion of devices in the network that are in that particular state. To model the coupled dynamics of this information diffusion process, we can make use of the SIS–SIS interaction model [120] from mathematical epidemiology. A state transition diagram is given by Figure 7.3. Notice that a change in status of one of the messages is allowed at each time instant. The set of differential equations describing the state evolution are as follows:

$$\frac{dUU_{k,l}(t)}{dt} = -(\alpha^{(1)}k\Theta_1 + \alpha^{(2)}l\Theta_2)UU_{k,l}(t) + IU_{k,l}(t) + UI_{k,l}(t), \tag{7.11}$$

$$\frac{dIU_{k,l}(t)}{dt} = \alpha^{(1)}k\Theta_1 UU_{k,l}(t) - (\alpha^{(2)}l\Theta_2 + 1)IU_{k,l}(t) + II_{k,l}(t), \tag{7.12}$$

Figure 7.3 State transition diagram for the simultaneous diffusion of two different messages in the IoBT network. At each time instant, a change in status of only one of the messages is allowed. The arrows are labeled with the transition probabilities.

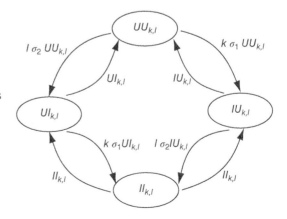

$$\frac{dUI_{k,l}(t)}{dt} = \alpha^{(2)}l\Theta_2 UU_{k,l}(t) - (\alpha^{(1)}k\Theta_1 + 1)UI_{k,l}(t) + II_{k,l}(t), \qquad (7.13)$$

$$\frac{dII_{k,l}(t)}{dt} = \alpha^{(1)}k\Theta_1 UI_{k,l}(t) + \alpha^{(2)}l\Theta_2 IU_{k,l}(t) - 2II_{k,l}(t), \qquad (7.14)$$

where Θ_1 and Θ_2 are the effective probabilities of a completely uninformed device, i.e. in state $UU_{k,l}$, to get informed by message 1 and message 2, respectively. They can be evaluated as follows:

$$\Theta_1 = \frac{1}{\mathbb{E}[K_1]} \sum_{k,l} \mathbb{P}(K_1 = k, K_2 = l)k(IU_{k,l,\sigma_1,\sigma_2}(t) + II_{k,l,\sigma_1,\sigma_2}(t)), \qquad (7.15)$$

$$\Theta_2 = \frac{1}{\mathbb{E}[K_2]} \sum_{k,l} \mathbb{P}(K_1 = k, K_2 = l)l(UI_{k,l,\sigma_1,\sigma_2}(t) + II_{k,l,\sigma_1,\sigma_2}(t)). \qquad (7.16)$$

Since $UU_{k,l}(t) + UI_{k,l}(t) + IU_{k,l}(t) + II_{k,l}(t) = 1$, in essence, only three of the four differential equations are linearly independent. The particular quantity of interest is $II_{k,l}(t)$ as it signifies the proportion of devices that are informed by both messages simultaneously at any particular time. This may be crucial in the IoBT setting where the devices require information of both layers to make accurate decisions. Note that the developed dynamics can be easily extended to the case of multiple message propagation, which will require a larger state space; however, in this section, we restrict ourselves to the case of only two network layers and messages.

7.3.2 Steady State Analysis

We are interested in determining the steady state or equilibrium of the information dissemination process since it characterizes the eventual spread of information in the network, which is independent of time. Although, with the changes in network topology and other network configurations, the actual information spread might be different; however, the equilibrium state provides us with a reasonable understanding of the system behavior. For the single message dissemination case, we impose the stationarity condition, i.e. set $\frac{dI_k^{(c)}(t)}{dt} = 0$. It results in the following expression:

$$I_k^{(c)*} = \frac{\alpha^{(c)}k\Theta^{(c)}(\alpha^{(c)})}{1 + \alpha^{(c)}k\Theta^{(c)}(\alpha^{(c)})}. \qquad (7.17)$$

Notice that $\Theta^{(c)}(\alpha^{(c)})$ is now a constant that depends on $\alpha^{(c)}$. Now, (7.10) and (7.17) present a system of equations that needs to be solved self-consistently to obtain the solution for $\Theta^{(c)}(\alpha^{(c)})$ and $I_k^{(c)*}$. For the multiple message propagation case, we can first write the reduced system in terms of the three independent states by substituting $UU_{k,l}(t) = 1 - IU_{k,l}(t) - UI_{k,l}(t) - II_{k,l}(t)$ as follows:

$$\frac{dIU_{k,l}(t)}{dt} = \alpha^{(1)}k\Theta_1 - (\alpha^{(1)}k\Theta_1 + \alpha^{(2)}l\Theta_2 + 1)IU_{k,l}(t)$$
$$- \alpha^{(1)}k\Theta_1 UI_{k,l}(t) - (\alpha^{(1)}k\Theta_1 - 1)II_{k,l}(t), \tag{7.18}$$

$$\frac{dUI_{k,l}(t)}{dt} = \alpha^{(2)}l\Theta_2 - \alpha^{(2)}l\Theta_2 IU_{k,l}(t) + (\alpha^{(1)}k\Theta_1 + \alpha^{(2)}l\Theta_2 + 1)UI_{k,l}(t)$$
$$- (\alpha^{(2)}l\Theta_2 - 1)II_{k,l}(t), \tag{7.19}$$

$$\frac{dII_{k,l}(t)}{dt} = \alpha^{(1)}k\Theta_1 UI_{k,l}(t) + \alpha^{(2)}l\Theta_2 IU_{k,l}(t) - 2II_{k,l}(t). \tag{7.20}$$

Using the stationarity condition, i.e. $\frac{dIU_{k,l}(t)}{dt} = 0$, $\frac{dUI_{k,l}(t)}{dt} = 0$, and $\frac{dII_{k,l}(t)}{dt} = 0$, we obtain the following expressions:

$$IU_{k,l}^* = \frac{\alpha^{(1)}k\Theta_1}{(1 + \alpha^{(1)}k\Theta_1)(1 + \alpha^{(2)}l\Theta_2)}, \tag{7.21}$$

$$UI_{k,l}^* = \frac{\alpha^{(2)}l\Theta_2}{(1 + \alpha^{(1)}k\Theta_1)(1 + \alpha^{(2)}l\Theta_2)}, \tag{7.22}$$

$$II_{k,l}^* = \left(\frac{\alpha^{(1)}k\Theta_1}{1 + \alpha^{(1)}k\Theta_1} \right) \left(\frac{\alpha^{(2)}l\Theta_2}{1 + \alpha^{(2)}l\Theta_2} \right), \tag{7.23}$$

Now, Eqs. (7.21)–(7.23) and Eqs. (7.15) and (7.16) need to be solved self-consistently to obtain the equilibrium solution of the quantity of interest, i.e. $II_{k,l}^*$. In the subsequent section, we present the methodology to obtain the solution to the dynamical information spreading process for the IoBT network and develop a framework that can assist in the planning and efficient design of such networks.

7.4 Mission-Specific Network Optimization

In this section, we first present a solution to the dynamical information spreading system developed in Section 7.3 and then use it for the efficient design of IoBT networks for mission-specific battlefield applications.

7.4.1 Equilibrium Solution

In order to find the equilibrium solution for the single network-wide message propagation case, we need to solve the self-consistent system expressed in

(7.10) and (7.17). In fact, it reduces to obtaining a solution to the following fixed-point system:

$$\Theta^{(c)}(\alpha^{(c)}) = \frac{1}{\mathbb{E}[K_c]} \sum_{k \geq 0} k \mathbb{P}(K_c = k) \frac{\alpha^{(c)} k \Theta^{(c)}(\alpha^{(c)})}{1 + \alpha^{(c)} k \Theta^{(c)}(\alpha^{(c)})}. \tag{7.24}$$

An obvious solution for this fixed-point system is $\Theta^{(c)}(\alpha^{(c)}) = 0$; however, it is noninformative. The existence of a nonzero solution is stated in the following theorem:

Theorem 7.1 *The fixed point equation in (7.24) relating to the information dissemination dynamics may have at least one solution in the domain $\Theta^{(c)}(\alpha^{(c)}) > 0$ depending on the value of $\alpha^{(c)}$. The condition for this bifurcation to hold is $\alpha^{(c)} \geq \frac{\mathbb{E}[K_c]}{\mathbb{E}[K_c^2]}$. This bifurcation point is unique in the domain $0 < \Theta^{(c)}(\alpha^{(c)}) \leq 1$.*

Proof: To prove that the fixed point equation described by (7.24) has a unique solution in the domain $\Theta^{(c)} > 0$, we make use of the Banach fixed-point theorem (or contraction mapping theorem) [63]. We prove that the functional

$$F(\Theta(\alpha^{(c)})) = \frac{1}{\mathbb{E}[K_c]} \mathbb{E}\left[\frac{K_c^2 \alpha^{(c)} \Theta(\alpha^{(c)})}{1 + K_c \alpha^{(c)} \Theta(\alpha^{(c)})}\right] \tag{7.25}$$

experiences a contraction for all $\Theta(\alpha^{(c)}) \in [0, 1]$. More precisely, we prove that $|F(\Theta_1(\alpha^{(c)})) - F(\Theta_2(\alpha^{(c)}))| \leq c|\Theta_1(\alpha^{(c)}) - \Theta_2(\alpha^{(c)})|$ for any $\Theta_1(\alpha^{(c)}), \Theta_2(\alpha^{(c)}) \in [0, 1]$, where $0 \leq c < 1$. The fact that the constant c is strictly less than 1 implies that the functional is contracted. The proof is as follows:

$$|F(\Theta_1(\alpha^{(c)})) - F(\Theta_2(\alpha^{(c)}))|$$

$$= \left|\frac{1}{\mathbb{E}[K_c]} \mathbb{E}\left[\frac{K_c^2 \alpha^{(c)} \Theta_1(\alpha^{(c)})}{1 + K_c \alpha^{(c)} \Theta_1(\alpha^{(c)})}\right] - \frac{1}{\mathbb{E}[K_c]} \mathbb{E}\left[\frac{K_c^2 \alpha^{(c)} \Theta_2(\alpha^{(c)})}{1 + K_c \alpha^{(c)} \Theta_2(\alpha^{(c)})}\right]\right|,$$

$$= \frac{|\Theta_1(\alpha^{(c)}) - \Theta_2(\alpha^{(c)})|}{\mathbb{E}[K_c]} \mathbb{E}\left[\frac{K_c^2 \alpha^{(c)}}{(1 + K_c \alpha^{(c)} \Theta_1(\alpha^{(c)}))(1 + K_c \alpha^{(c)} \Theta_2(\alpha^{(c)}))}\right]. \tag{7.26}$$

To complete the proof, we need to show that

$$\frac{1}{\mathbb{E}[K_c]} \mathbb{E}\left[\frac{K_c^2 \alpha^{(c)}}{(1 + K_c \alpha^{(c)} \Theta_1(\alpha^{(c)}))(1 + K_c \alpha^{(c)} \Theta_2(\alpha^{(c)}))}\right] < 1, \tag{7.27}$$

Let $g(K_c) = \frac{K_c^2 \alpha^{(c)}}{(1 + K_c \alpha^{(c)} \Theta_1(\alpha^{(c)}))(1 + K_c \alpha^{(c)} \Theta_2(\alpha^{(c)}))}$. It can be proved that $g(K_c)$ is concave for $K_c \geq 0$ by showing that $g''(K_c) < 0, \forall K_c \geq 0$. Therefore, using Jensen's inequality [27], we can say that $\mathbb{E}[g(K_c)] \leq g(\mathbb{E}[K_c])$, with equality iff K_c is deterministic.

It follows that

$$\frac{1}{\mathbb{E}[K_c]} \mathbb{E}\left[\frac{K_c^2 \alpha^{(c)}}{(1 + K_c \alpha^{(c)} \Theta_1(\alpha^{(c)}))(1 + K_c \alpha^{(c)} \Theta_2(\alpha^{(c)}))} \right]$$

$$\leq \frac{\mathbb{E}[K_c]\alpha^{(c)}}{(1 + \mathbb{E}[K_c]\alpha^{(c)}\Theta_1(\alpha^{(c)}))(1 + \mathbb{E}[K_c]\alpha^{(c)}\Theta_2(\alpha^{(c)}))}$$

$$= \frac{1}{\Theta_1(\alpha^{(c)}) + \Theta_2(\alpha^{(c)}) + \mathbb{E}[K_c]\alpha^{(c)}\Theta_1(\alpha^{(c)})\Theta_2(\alpha^{(c)}) + \frac{1}{\mathbb{E}[K_c]\alpha^{(c)}}}. \qquad (7.28)$$

The expression in (7.28) is strictly less than 1 only if the following condition is satisfied:

$$\Theta_1(\alpha^{(c)}) + \Theta_2(\alpha^{(c)}) + \mathbb{E}[K_c]\alpha^{(c)}\Theta_1(\alpha^{(c)})\Theta_2(\alpha^{(c)}) + \frac{1}{\mathbb{E}[K_c]\alpha^{(c)}} > 1. \qquad (7.29)$$

The condition in (7.29) depends on the relative magnitudes of the quantities $\mathbb{E}[K_c]$ and $\alpha^{(c)}$. Regardless, it reveals that we need to exclude the values of $\Theta(\alpha^{(c)})$ that are too close to zero. For sufficiently large values of $\Theta(\alpha^{(c)})$, it is clear from (7.29) that $F(\Theta(\alpha^{(c)}))$ is indeed a contraction with respect to the absolute value metric. Hence, by the contraction mapping theorem, $F(\Theta(\alpha^{(c)}))$ has a unique fixed point in the domain $\Theta(\alpha^{(c)}) > 0$. The nonzero equilibrium solution can be obtained by solving the following equation:

$$1 = \frac{1}{\mathbb{E}[K_c]} \mathbb{E}\left[\frac{K_c^2 \alpha^{(c)}}{1 + K_c \alpha^{(c)} \Theta(\alpha^{(c)})} \right]. \qquad (7.30)$$

Let $h(\Theta(\alpha^{(c)})) = \frac{1}{\mathbb{E}[K_c]} \mathbb{E}\left[\frac{K_c^2 \alpha^{(c)}}{1 + K_c \alpha^{(c)} \Theta(\alpha^{(c)})} \right]$. We need to find a solution to the equation $h(\Theta(\alpha^{(c)})) = 1$ in the domain $0 < \Theta(\alpha^{(c)}) \leq 1$. It is clear that $h(\Theta(\alpha^{(c)}))$ is monotonically decreasing for $\Theta(\alpha^{(c)}) > 0$. Therefore, it is sufficient to show that $h(0) > 1$ and $h(1) < 1$ for a unique nonzero solution to exist for the equation $h(\Theta(\alpha^{(c)})) = 1$. This result is proved as follows:

$$h(0) = \frac{1}{\mathbb{E}[K_c]} \mathbb{E}[K_c^2 \alpha^{(c)}] = \alpha^{(c)} \frac{\mathbb{E}[K_c^2]}{\mathbb{E}[K_c]}, \qquad (7.31)$$

$$h(1) = \frac{1}{\mathbb{E}[K_c]} \mathbb{E}\left[\frac{K_c^2 \alpha^{(c)}}{1 + K_c \alpha^{(c)}} \right] = \frac{1}{\mathbb{E}[K_c]} \mathbb{E}\left[K_i \frac{K_c \alpha^{(c)}}{1 + K_c \alpha^{(c)}} \right]$$

$$< \frac{1}{\mathbb{E}[K_c]} \mathbb{E}[K_c] = 1. \qquad (7.32)$$

In (7.32), the inequality follows from the fact that $\frac{K_c \alpha^{(c)}}{1 + K_c \alpha^{(c)}} < 1, \forall K_c > 0, \alpha^{(c)} > 0$.

A nonzero solution to (7.30) exists only if $h(0) \geq 1$, which implies that $\alpha^{(c)} \geq \frac{\mathbb{E}[K_c]}{\mathbb{E}[K_c^2]}$. This is exactly the critical spreading rate, also known as epidemic threshold, in the SIS model [110]. $\qquad \square$

Obtaining a closed form solution for the fixed point system in (7.24) for a PPP setting is not possible due to the complex form of $\mathbb{P}(K_c = k)$. Hence, an approximate solution can be obtained using the following theorem:

Theorem 7.2 *If a nonzero solution exists for the information spreading dynamics in (7.10) and (7.17), i.e. $\alpha^{(c)} \geq \frac{\mathbb{E}[K_c]}{\mathbb{E}[K_c^2]}$, then for $\mathbb{E}[K_c] \gg 1$, a lower bound approximation of the solution can be expressed as follows:*

$$\Theta^{(c)}(\alpha^{(c)}) \approx \left(1 - \frac{1}{\alpha^{(c)}\mathbb{E}[K_c]}\right)^{+}, \tag{7.33}$$

where $(x)^{+}$ represents $\max(0, x)$.

Proof: Obtaining the nonzero solution for the fixed point Eq. (7.24) in closed form is not possible since we need to solve the following equation for $\Theta(\alpha^{(c)})$:

$$1 = \frac{1}{\mathbb{E}[K_c]} \sum_{k=0}^{\infty} \frac{K_c^2 \alpha^{(c)}}{1 + K_c \alpha^{(c)}\Theta(\alpha^{(c)})} \mathbb{P}(K_c = k), \tag{7.34}$$

where $P(K_c = k)$ is difficult to obtain in closed form. Therefore, we resort to finding an approximation for the solution that is asymptotically accurate. Let $g(K_c) = \frac{K_c^2 \alpha^{(c)}}{1 + K_c \alpha^{(c)}\Theta(\alpha^{(c)})}$. Since $g''(K_c) > 0, \forall K_i \geq 0$, so $g(K_i)$ is a convex function for $K_i \geq 0$. Using Jensen's inequality, we can say that $\mathbb{E}[g(K_c)] \geq g(\mathbb{E}[K_c])$, with equality only if K_c is deterministic. This implies the following:

$$\mathbb{E}\left[\frac{K_c^2 \alpha^{(c)}}{1 + K_c \alpha^{(c)}\Theta(\alpha^{(c)})}\right] \geq \frac{\mathbb{E}[K_c]^2 \alpha^{(c)}}{1 + \mathbb{E}[K_c]\alpha^{(c)}\Theta(\alpha^{(c)})}. \tag{7.35}$$

Therefore, we can write (7.34) as follows:

$$1 \geq \frac{\mathbb{E}[K_c]\alpha^{(c)}}{1 + \mathbb{E}[K_c]\alpha^{(c)}\Theta(\alpha^{(c)})}, \tag{7.36}$$

which leads to the final solution,

$$\Theta(\alpha^{(c)}) \geq 1 - \frac{1}{\alpha^{(c)}\mathbb{E}[K_c]}. \tag{7.37}$$

Using our prior knowledge that $\Theta(\alpha^{(c)}) \geq 0$, we need to ensure that $\alpha^{(c)}\mathbb{E}[K_c] \geq 1$. In general, the complete solution can be expressed as $\Theta(\alpha^{(c)}) \geq \max(0, 1 - \frac{1}{\alpha^{(c)}\mathbb{E}[K_c]})$. To measure the accuracy of this bound, we solve the fixed-point equation exactly using a fixed-point iteration and compare the results for different values of $\alpha^{(c)}$ and $\mathbb{E}[K_c] = \lambda_i \pi r_i^2$. We choose a fixed $r_i = 0.2\,\text{km}$ and $\lambda_i = [25, 50, 100]\,\text{km}^{-2}$, which results in $\mathbb{E}[K_i] = [3.14, 6.28, 12.57]$. A plot of the results is provided in Figure 7.4. It can be observed that the lower bound obtained using Jensen's inequality is tight for all values of $\alpha^{(c)}$ when $\mathbb{E}[K_i] \gg 1$. $\qquad\square$

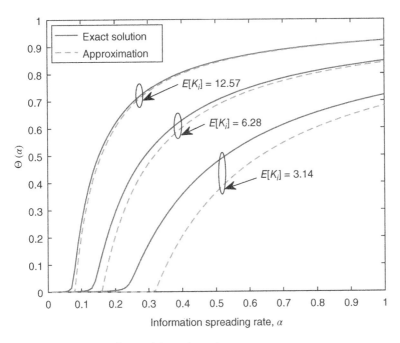

Figure 7.4 Accuracy of Jensen's lower bound.

Theorem 7.2 provides a lower bound for the exact solution and becomes a tight approximation for $\mathbb{E}[K_c] \gg 1$. Moreover, the approximation has also resulted in an increase of the critical information spreading threshold to $\alpha^{(c)} \geq \frac{1}{\mathbb{E}[K_c]}$ to ensure that $\Theta^{(c)} \geq 0$. This is also asymptotically accurate as the original condition can be written as $\alpha^{(c)} \geq \frac{\mathbb{E}[K_c]}{\mathbb{E}[K_c^2]} = \frac{1}{\mathbb{E}[K_c] + \frac{\sigma_{K_c}^2}{\mathbb{E}[K_c]}}$, where $\sigma_{K_c}^2$ is the variance of K_c. It approaches $\frac{1}{\mathbb{E}[K_c]}$ as $\mathbb{E}[K_c]$ becomes large. The relaxed condition is formally expressed by the following corollary.

Corollary 7.1 *The condition for obtaining an approximate nonzero equilibrium for the information dissemination dynamics is given as follows:*

$$\alpha^{(c)} \geq \frac{1}{\mathbb{E}[K_c]}. \tag{7.38}$$

In the IoBT network, the physical interpretation of $\mathbb{E}[K_c]$ is that the average number of communication neighbors of a device in the combined network and hence, it is reasonable to assume that $\mathbb{E}[K_c] \geq 1$ due to the potential high density of devices in IoBT networks. Therefore, the solution presented in Theorem 7.2 is indeed a good approximation to the actual solution. The corresponding solution for the density of informed devices can be obtained using (7.17).

In the case of multiple message propagation, we need to solve the self-consistent system of equations defined by Eqs. (7.21)–(7.23) and Eqs. (7.15) and (7.16). It reduces to solving the following fixed-point equations:

$$\Theta_1 = \frac{1}{\mathbb{E}[K_1]} \sum_{k,l} \mathbb{P}(K_1 = k, K_2 = l) k \left(\frac{\alpha^{(1)} k \Theta_1 + \alpha^{(1)^2} k l \Theta_1 \Theta_2}{(1 + \alpha^{(1)} k \Theta_1)(1 + \alpha^{(2)} l \Theta_2)} \right),$$

$$= \frac{1}{\mathbb{E}[K_1]} \sum_{k} \mathbb{P}(K_1 = k) k \left(\frac{\alpha^{(1)} k \Theta_1}{1 + \alpha^{(1)} k \Theta_1} \right), \tag{7.39}$$

$$\Theta_2 = \frac{1}{\mathbb{E}[K_2]} \sum_{k,l} \mathbb{P}(K_1 = k, K_2 = l) l \left(\frac{\alpha^{(2)} l \Theta_2 + \alpha^{(2)^2} k l \Theta_1 \Theta_2}{(1 + \alpha^{(1)} k \Theta_1)(1 + \alpha^{(2)} l \Theta_2)} \right),$$

$$= \frac{1}{\mathbb{E}[K_2]} \sum_{l} \mathbb{P}(K_2 = l) l \left(\frac{\alpha^{(2)} l \Theta_2}{1 + \alpha^{(2)} l \Theta_2} \right). \tag{7.40}$$

Notice that Θ_1 and Θ_2 are independent of each other and their fixed point equations are similar to (7.24). Hence, the existence and uniqueness of the fixed point can be proved under similar conditions. Using a similar approach to the solution of the single message propagation case, a lower bound approximate solution is provided by the following theorem:

Theorem 7.3 *If a nonzero solution exists for the information spreading dynamics for the case two simultaneous message propagation, i.e. $\alpha^{(1)} \geq \frac{\mathbb{E}[K_1]}{\mathbb{E}[K_1^2]}$ and $\alpha^{(2)} \geq \frac{\mathbb{E}[K_2]}{\mathbb{E}[K_2^2]}$, then for $\mathbb{E}[K_1] \gg 1$ and $\mathbb{E}[K_2] \gg 1$, a lower bound approxima- tion of the solution can be expressed as follows:*

$$\Theta_1(\alpha) \approx \left(1 - \frac{1}{\alpha^{(1)} \mathbb{E}[K_1]} \right)^+, \tag{7.41}$$

$$\Theta_2(\alpha) \approx \left(1 - \frac{1}{\alpha^{(2)} \mathbb{E}[K_2]} \right)^+. \tag{7.42}$$

Proof: The objective is to obtain a fixed point solution to the equations (7.39) and (7.40). Since the equations are decoupled, we can employ the same approach used for Theorem 7.2 to show the existence and uniqueness of the fixed point. Consequently, a lower bound approximation of the solution for each fixed-point equation can be obtained using Jensen's inequality. The condition for existence of a solution is $\alpha^{(c)} \geq \frac{\mathbb{E}[K_1]}{\mathbb{E}[K_1^2]}$ and $\alpha^{(c)} \geq \frac{\mathbb{E}[K_2]}{\mathbb{E}[K_2^2]}$, which can be written as $\alpha^{(c)} \geq \max \left(\frac{\mathbb{E}[K_1]}{\mathbb{E}[K_1^2]}, \frac{\mathbb{E}[K_2]}{\mathbb{E}[K_2^2]} \right)$. □

Similar to Corollary 7.1, the condition for obtaining an approximate nonzero equilibrium solution is $\alpha^{(i)} \geq 1/\mathbb{E}[K_i], i \in \{1, 2\}$. The corresponding solution for the proportion of devices that are informed about both messages, i.e. $II_{k,l}$, can be obtained using (7.23), which turns out to be the following:

$$II^*_{k,l} = I_k^{(1)*} \times I_l^{(2)*}. \tag{7.43}$$

This result is interesting and useful as it can be easily generalized for the case of multiple connectivity layers, i.e. types of devices, and multiple messages propagating simultaneously.

7.4.2 Secure and Reconfigurable Network Design

Once the equilibrium point for information dissemination has been determined, the next step is to design the IoBT network to achieve mission specific goals while efficiently using battlefield resources. In essence, the network design implies tuning the knobs of the network, which in the case of IoBT networks are the transmission ranges and the node deployment densities of the different types of battlefield things. However, changing the physical parameters may have an impact on the cost and hence, the goal is to ensure a certain information spreading profile in the network while deploying the minimum number of devices and using the minimum transmit power. Let $\mathbf{r} = [r_1 \; r_2]^T$ represent the vector of communication ranges of each of the type-I and type-II devices in the IoBT network. The minimum density of the devices in the network, determined by the mission requirements, is denoted by $\lambda^{\min}$, where $\lambda^{\min} \geq 0$. The maximum deployment density of the devices, defined by the capacity of the available devices, is denoted by $\lambda^{\max}$, where $\lambda^{\max} \geq \lambda^{\min}$. Similarly, the tunable transmission range limits of the devices can be expressed as $\mathbf{r}^{\min} = [r_1^{\min} \; r_2^{\min}]^T$, $r_m^{\min} \geq 0$, $\forall m \in \{1,2\}$, and $\mathbf{r}^{\max} = [r_1^{\max} \; r_2^{\max}]^T$, $r_m^{\max} \geq r_m^{\min}$, $m \in \{1,2\}$. If $\mathbf{w} = [w_1 \; w_2]^T$, such that $\sum_{m=1}^2 w_m = 1$ represents the weight vector corresponding to the relative capital cost of deploying a type-I and type-II device, respectively, and c represents the unit operational power cost signifying the importance of network power consumption, then a cost function for the network with device densities λ and transmission ranges $\mathbf{r}$ can be expressed as follows:

$$C(p, \lambda, \mathbf{r}) = w_1 p \lambda + w_2 (1-p) \lambda + c(p \lambda r_1^\eta + \lambda r_2^\eta), \tag{7.44}$$

where η denotes the path-loss exponent.[4] The first term represents the total deployment cost per unit area of all the network devices, while the second term represents the total energy cost per unit area of operating all the devices with transmission range $\mathbf{r}$. The weights $\mathbf{w}$ can depend on several factors such as the time required for deployment, the monetary cost involved, or the number of devices available in stock, etc. We can then formulate the secure

---

4 The power consumption of a device with communication range r_m is proportional to r_m^η.

and reconfigurable network design problem as follows:

$$\underset{p,\lambda,\mathbf{r}}{\text{minimize}} \quad C(p,\lambda,\mathbf{r}) \tag{7.45}$$

$$\text{subject to} \quad I_k^{(c)*} \geq T_k^{(c)}, \forall k \geq 0, \tag{7.46}$$

$$II_{k,l}^* \geq T_{k,l}, \forall k \geq 0, l \geq 0, \tag{7.47}$$

$$p^{\min} \leq p \leq p^{\max}, \lambda^{\min} \leq \lambda \leq \lambda^{\max}, \mathbf{r}^{\min} \leq \mathbf{r} \leq \mathbf{r}^{\max}, \tag{7.48}$$

where $T_k^{(c)} \in (0,1)$, $k \geq 0$ are the minimum desired proportions of degree k devices that are informed with a single network-wide message, $T_{k,l} \in (0,1)$, $k \geq 0, l \geq 0$ are the desired proportion of devices with degree $K_1 = k$ and $K_2 = l$ that are informed with both messages simultaneously, and $p^{\min}$ and $p^{\max}$ are the minimum and maximum fractions of devices, respectively, that are of type-I. The cost function C is a convex nondecreasing function of p, λ, and $\mathbf{r}$. However, it is an infinite dimension optimization problem due to a constraint on each degree class of devices. To be able to solve this problem, we need to select a desired mapping $T_k^{(c)}$ for the combined degree $K_c = k$ and similarly a mapping $T_{k,l}$ for devices with joint intra-layer degrees $K_1 = k, K_2 = l, \forall k, l \geq 0$. Further investigation reveals that for a fixed α, the information dissemination can only have restricted trajectories based on the average device degree as shown in Figure 7.5. Hence, the threshold mappings cannot be defined arbitrarily as they might not be achievable. To handle this problem, we can express the constraints in (7.46), assuming that $\alpha^{(c)}$ satisfies the condition in Corollary 7.1, as follows:

$$\mathbb{E}[K_c] \geq \frac{1}{\alpha^{(c)} - T_k^{(c)}/(k(1 - T_k^{(c)}))}, \quad \forall k \geq 0. \tag{7.49}$$

Now, we need to satisfy (7.49) for each $(k, T_k^{(c)})$ pair, and from Figure 7.5, it is clear that there is no incentive to choose a threshold profile that is different from one of the possible trajectories. Therefore, specifying the threshold for a single value of the degree is sufficient to characterize the entire trajectory. Since K_c is a random variable with a distribution centered around $\mathbb{E}[K_c]$, it is plausible to set a threshold on the proportion of devices with the average device degree, i.e. using the pair $(\mathbb{E}[K_c], T_{\mathbb{E}[K_c]}^{(c)})$. It results in the following single constraint instead of the infinite set of constraints in (7.46):

$$\mathbb{E}[K_c] \geq \frac{1}{\alpha^{(c)}(1 - T_{\mathbb{E}[K_c]}^{(c)})}. \tag{7.50}$$

Note that this constraint implies $\alpha^{(c)}\mathbb{E}[K_c] \geq 1/(1 - T_{\mathbb{E}[K_c]}^{(c)})$, which satisfies the condition in Corollary 7.1, $\forall T_{\mathbb{E}[K_c]}^{(c)} \in [0,1]$, thus validating our assumption. For

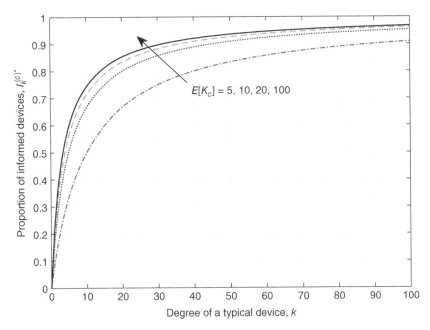

Figure 7.5 Information dissemination profiles for varying average degree of devices ($\alpha = 0.3$).

the second set of constraints in (7.47), due to the decomposability of $II_{k,l}^*$ as shown in (7.43) and the fact that $I_k^{(i)*} \in [0,1], i \in \{1,2\}$, the constraints can be separated as follows:

$$\frac{\alpha^{(1)}k\Theta_1}{1+\alpha^{(1)}k\Theta_1} \geq T_{k,l}, \quad \frac{\alpha^{(2)}l\Theta_2}{1+\alpha^{(2)}l\Theta_2} \geq T_{k,l}, \quad \forall k,l \geq 0. \tag{7.51}$$

Moreover, the thresholds can be replaced by simply T_k and T_l instead of $T_{k,l}$ as they are constants. Using a similar approach as before, we replace the set of infinite constraints in (7.47) by the following two constraints:

$$\mathbb{E}[K_i] \geq \frac{1}{\alpha^{(i)}(1-T_{\mathbb{E}[K_i]})}, \quad i \in \{1,2\}. \tag{7.52}$$

For brevity, we henceforth denote $T_{\mathbb{E}[K_c]}^{(c)}$, $T_{\mathbb{E}[K_1]}$, and $T_{\mathbb{E}[K_2]}$ as simply $T^{(c)}$, $T^{(1)}$, and $T^{(2)}$. Furthermore, we denote $\frac{1}{\alpha^{(c)}(1-T^{(c)})}$, $\frac{1}{\alpha^{(1)}(1-T^{(1)})}$, and $\frac{1}{\alpha^{(2)}(1-T^{(2)})}$ by $\mathcal{T}^{(c)}$, $\mathcal{T}^{(1)}$, and $\mathcal{T}^{(2)}$, respectively, corresponding to the desired minimum success probabilities $\mathcal{P}_s^{(i)}$ leading to $\alpha^i, i \in \{1,2,c\}$. The original optimization problem can then be rewritten as follows:

$$\underset{p,\lambda,\mathbf{r}}{\text{minimize}} \quad w_1 p\lambda + w_2(1-p)\lambda + c(p\lambda r_1^\eta + \lambda r_2^\eta), \tag{7.53}$$

$$\text{subject to} \quad p^2 \lambda \pi r_1^2 + \lambda \pi r_2^2 \geq \mathcal{T}^{(c)}, \tag{7.54}$$

$$p^2 \lambda \pi r_1^2 \geq \mathcal{T}^{(1)}, \tag{7.55}$$

$$\lambda \pi r_2^2 \geq \mathcal{T}^{(2)}, \tag{7.56}$$

$$p^{\min} \leq p \leq p^{\max}, \lambda^{\min} \leq \lambda \leq \lambda^{\max}, \mathbf{r}^{\min} \leq \mathbf{r} \leq \mathbf{r}^{\max}. \tag{7.57}$$

It is important to note that the conditions for existence of nonzero equilibrium in Corollary 7.1 and the ones resulting from Theorem 7.3 are implicitly incorporated into the constraints and do not need to be imposed separately. This implies that if a feasible solution to the optimization problem exists, then there exists a nonzero equilibrium solution to the information dissemination dynamics. The objective and constraints are nondecreasing smooth functions of the optimization variables. It is clear that the objective and constraints are convex in the feasible solution space. Therefore, the problem can be solved using standard convex optimization techniques [140].

The secure and resilient framework for mission critical information dissemination in IoBT networks is provided in Algorithm 7.1. At the beginning of the mission, the central network planner obtains the optimal physical network parameters by solving the optimization problem in Eqs. (7.53)–(7.57), and accordingly deploys the devices with appropriate transmission powers. The central planner then periodically analyzes the connectivity situation of the network with a reconfigurability interval denoted by t_r. The reconfigurability interval could range from several hours to days depending on the mission requirements. Several techniques may be employed to estimate the connectivity, and consequently the information dissemination in the network. The effect of physical damage can be measured by physically monitoring the network with the aid of robotic systems such as in [113]. On the other hand, the effect of cyber attacks may be estimated by running discovery tests on the network to assess the reachability of nodes. Based on these results, estimates of the information dissemination thresholds can be obtained, i.e. $\hat{\mathcal{T}}^{(1)}$, $\hat{\mathcal{T}}^{(2)}$, and $\hat{\mathcal{T}}^{(c)}$. If there is a significant drop in the estimated threshold as compared with the desired one or there is a change in the required security level, then the optimization needs to be re-computed. The new optimized parameters help in identifying the additional deployment needed for each type of devices and/or the reconfiguration of their transmission powers to achieve the desired information dissemination in the IoBT network. In Section 7.3, we investigate the behavior of the optimal solutions under varying threat levels and mission specific performance thresholds.

Algorithm 7.1 Secure and Reconfigurable Network Design

1: At epoch, i.e. $t = 0$; Initialize requirements for information dissemination, i.e. $T^{(1)}$, $T^{(2)}$, and $T^{(c)}$ and the anticipated threat level δ.

2: Obtain optimal network parameters $\lambda_1^{\text{init}} = p^{\text{init}}\lambda^{\text{init}}$, $\lambda_2^{\text{init}} = \lambda^{\text{init}}, r_1^{\text{init}}$, and r_2^{init} by solving the optimization problem in Eqs. (7.53) to (7.57) and accordingly deploy the devices with the appropriate communication ranges.

3: **repeat**

4:     **if** $t = \zeta t_r, \zeta \in \mathbb{Z}^+$ **then**

5:         Obtain an estimate of the density of active devices $\hat{\lambda} = [\hat{\lambda}_1, \hat{\lambda}_2]$ and use the initial communication ranges to estimate the prevailing information dissemination level $\hat{T}^{(1)}$, $\hat{T}^{(2)}$, and $\hat{T}^{(c)}$.

6:         Re-evaluate the desired security level in response to the threats and accordingly update the parameter $\hat{\delta}$.

7:         **if** $|T^{(1)} - \hat{T}^{(1)}| \geq \epsilon$ **or** $|T^{(2)} - \hat{T}^{(2)}| \geq \epsilon$ **or** $|T^{(c)} - \hat{T}^{(c)}| \geq \epsilon$ **or** $\hat{\delta} \neq \delta$, **then**

8:             Recompute the optimization problem in Eqs. (7.53) to (7.57) to obtain the new set of optimal parameters $\lambda_1^{\text{new}} \leftarrow p^{\text{new}}\lambda^{\text{new}}$, $\lambda_2^{\text{new}} \leftarrow \lambda^{\text{new}}$, r_1^{new}, and r_2^{new}.

9:             Deploy additional required $\lambda^{\text{new}} - \hat{\lambda}$ devices in the network and reconfigure transmission powers to achieve the required transmission ranges $\mathbf{r}^{\text{new}}$.

10:         **end if**

11:     **end if**

12:     $t \leftarrow t + 1$;

13: **until** End of mission.

7.5 Simulation Results and Validation

In this section, we provide the results obtained by testing the developed framework under different battlefield missions. We assume a bi-layer IoBT network comprising of type-I and type-II devices. The first type of devices is assumed to be commanders and the second type is assumed to be followers. The assumption yields a simple yet natural network configuration in a battlefield, e.g. being composed of soldiers and distributed commanding units. The allowable physical parameter ranges of the respective devices are selected to be as follows: the minimum and maximum device deployment density is selected as $\lambda^{\text{min}} = 1\text{km}^{-2}$ and $\lambda^{\text{max}} = 15\,\text{km}^{-2}$, respectively, the minimum and maximum fraction of type-I devices $p^{\text{min}} = 0$ and $p^{\text{max}} = 0.4$, respectively, the minimum and maximum communication ranges of devices in the first layer, $r_1^{\text{min}} = 100\,\text{m}$ and $r_1^{\text{max}} = 2000\,\text{m}$, respectively, and the minimum and maximum communication ranges of devices in the second layer,

$r_2^{\min} = 10$ m and $r_2^{\max} = 800$ m. The parameters imply that the active devices in the first layer, i.e. the commanding units, have a higher allowable transmission range than the followers. In practice, the limits can be based on tactical requirements of the missions. For simplicity, we assume that sufficient number of channels are available and for the considered densities and communication ranges of the devices, the MAC protocol is able to effectively mitigate interference in the communication resulting in a constant success probability. Further, we assume that the desired success probabilities $\mathcal{P}_s^{(i)} = 1, \forall i \in \{1, 2, c\}$ implying perfect success of transmissions. Consequently, $\alpha^{(1)} = \alpha^{(2)} = \alpha^{(c)}$ and is henceforth referred to as α. The weights representing the relative deployment cost are chosen to be $w_1 = 100$ and $w_2 = 50$, which implies that the deployment cost of the commanding units is twice as much as the follower units. The unit cost of power is selected to be $c = 100$ that can be adjusted according to the importance of each mission and the path-loss exponent $\eta = 4$.

7.5.1 Mission Scenarios

In the battlefields, there can be several types of missions such as intelligence, surveillance, encounter battle, espionage, reconnaissance, etc. In our results, we will focus particularly on the two most common mission scenarios, i.e. intelligence and encounter battle. Both of them have completely different requirements in terms of the information flow in the network, which are described as follows:

7.5.1.1 Intelligence

In the intelligence mission, the goal is to provide commanders with information from a range of sources to assist them in operational or campaign planning. This requires a high network-wide dissemination of a single message, while the condition for simultaneously informed devices may not be stringent. In essence, the commander network must reinforce the follower network to achieve a high network-wide message propagation. Hence, to emulate such an intelligence mission, we select the following set of information spreading thresholds: $T^{(1)} = T^{(2)} = 0.6$, $T^{(c)} = 0.8$. The optimal physical parameters obtained for the intelligence mission against increasing threat level δ are shown in Figure 7.6. There are several interesting observations in the intelligence mission. A general trend is that the required transmission ranges and deployment densities increases as the threat level increases. Consequently, the cost function, which signifies the deployment and operation cost of the network, also increases as shown in Figure 7.6c. Figure 7.6a,b shows that the transmission range of the commanders is always higher than the followers while the densities of the followers is higher than that of the commanders. This observation makes sense as the followers equipped with sensors should be

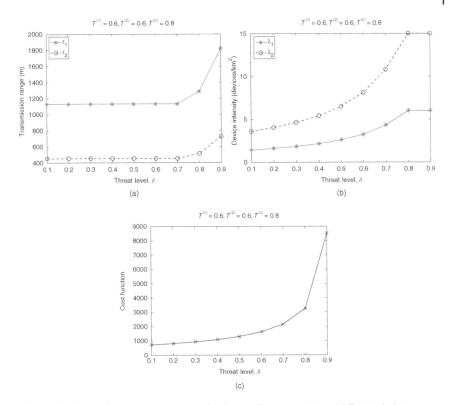

Figure 7.6 Optimal network parameters for the intelligence mission. (a) Transmission ranges of devices against varying threat level. (b) Deployment density of devices against varying threat level. (c) Cost function against varying threat level.

more in the total number while the commander network should have a larger influence area to be able to gather information for the intelligence mission. Another important observation is that the framework tends to increase the deployment density of the devices first before increasing their transmission ranges. It is due to a high cost of power consumption that tends to force the devices to minimize the transmission ranges.

7.5.1.2 Encounter Battle

In the encounter battle or meeting engagement scenario, there is a contact between the battling forces. In such situations, commanders need to act quickly to gain advantage over the opponents. This requires devices to be informed simultaneously about the information disseminated in both network layers. Hence, there is a need for strong coordination among commanders as well as followers. Additionally, the common status information sharing between all network devices must be sufficiently high to ensure accurate

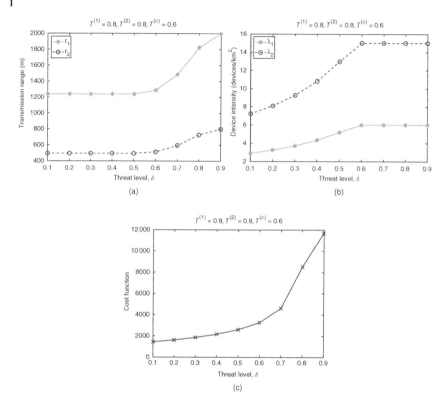

Figure 7.7 Optimal network parameters for the encounter mission. (a) Transmission ranges of devices against varying threat level. (b) Deployment density of devices against varying threat level. (c) Cost function against varying threat level.

decision-making. Therefore, we set the following information spreading thresholds: $T^{(1)} = T^{(2)} = 0.8$ and $T^{(c)} = 0.6$. The resulting optimal parameters against the changing threat level are presented in Figure 7.7. In contrast with the intelligence mission, the framework utilizes all the available devices at a much lower threat level and is also forced to increase the transmission ranges to full capacity despite the high power cost. The cost function for the encounter battle in Figure 7.7c is significantly higher than the intelligence mission in Figure 7.7c due to more stringent information spreading criteria, which requires more transmission power and higher device density.

Next, we study the impact of the information spreading thresholds on the optimal network parameters. In Figure 7.8, we fix the intra-layer information dissemination threshold to $T^{(1)} = T^{(2)} = 0.5$ and observe the change in parameters for varying network-wide information dissemination threshold $T^{(c)}$. Notice that the curves in Figure 7.8a,b are mostly flat except for very high

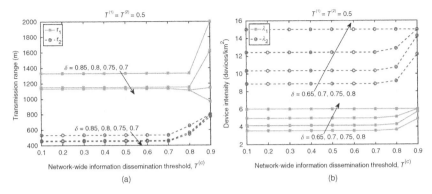

Figure 7.8 Network parameter variation with changes in combined-layer information dissemination threshold. (a) Transmission ranges of devices against varying network-wide information dissemination threshold. (b) Deployment density of devices against varying network-wide information dissemination threshold.

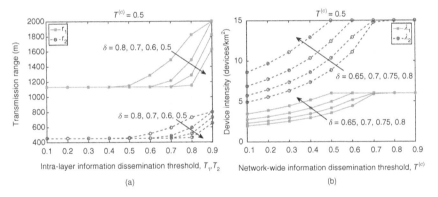

Figure 7.9 Network parameter variation with changes in intra-layer information dissemination threshold. (a) Transmission ranges of devices against varying intra-layer information dissemination threshold. (b) Deployment density of devices against varying intra-layer information dissemination threshold.

values of $T^{(c)}$. This implies that there is no need for additional resources to achieve a higher network-wide information dissemination, which reaffirms the fact that the intra-layer information dissemination is a much stricter condition than the network-wide information dissemination. Similarly, in Figure 7.9, we fix the network-wide information dissemination threshold to $T^{(c)} = 0.5$ and observe the change in optimal parameters with a change in intra-layer information dissemination threshold. In this case, the device densities are increased successively followed by the communication ranges as the threshold for the proportion of simultaneously informed devices is increased.

7.6   Summary and Conclusion

In this chapter, we have presented a generic framework for secure and reconfigurable design of IoT empowered battlefield networks. The framework provides a tractable approach to tune the physical network parameters to achieve the desired real-time data dissemination among different types of battlefield devices according to the assigned missions. It takes into account the perceived threat level from the opponent as well as the costs involved in the deployment and operation of combat equipment to provide a robust and cost effective design of communication networks in battlefields, which can be highly useful in military planning. Optimized network parameters are provided for the two typical mission scenarios of intelligence and encounter battle in which the desired information spreading objectives are different from one another. Results show that the mission goals can be achieved by either appropriately changing the deployment density of combat units or by changing their transmission powers or both in response to changing mission requirements. Moreover, the network can be reconfigured according to the periodic assessment of lost connectivity or changes in security requirements to meet the desired mission goals.

8

Network Defense Mechanisms Against Malware Infiltration

8.1 Malware Infiltration and Botnets

Malware may infect a large number of network devices using device-to-device (D2D) communication resulting in the formation of a botnet, i.e. a network of infected devices controlled by a common malware. A botmaster may exploit it to launch a network-wide attack sabotaging infrastructure and facilities, or for malicious purposes such as collecting ransom. In this chapter, we propose an analytical model to study the D2D propagation of malware in wireless Internet of things (IoT) networks. Leveraging tools from dynamic population processes and point process theory, we capture malware infiltration and coordination process over a network topology. The analysis of mean-field equilibrium in the population is used to construct and solve an optimization problem for the network defender to prevent botnet formation by patching devices while causing minimum overhead to network operation. The developed analytical model serves as a basis for assisting the planning, design, and defense of such networks from a defender's standpoint.

8.1.1 Network Model

We consider a set of wireless IoT devices uniformly distributed in $\mathbb{R}^2$ according to a homogeneous Poisson point process (PPP) [133] denoted by $\Phi = \{x_i\}_{i \geq 1}$ with intensity $\lambda \in \mathbb{N}$ devices/km^2, where $x_i \in \mathbb{R}^2$ represents the location coordinates of the ith device. Each device has computing capabilities for executing processes and has a wireless interface for communication with neighboring devices. The devices are assumed to have omni-directional transmissions with a communication range of r m. A typical device located at x_i is connected wirelessly with $K = |N_i|$ other devices, where $N_i = \{j : \|x_i - x_j\| \leq r, \forall j \neq i\}$ and $|\cdot|$ denotes the cardinality operator. Since the devices in the network are distributed according to a PPP, the degree K is a random variable with

Resource Management for On-Demand Mission-Critical Internet of Things Applications, First Edition.
Junaid Farooq and Quanyan Zhu.
© 2021 John Wiley & Sons, Inc. Published 2021 by John Wiley & Sons, Inc.

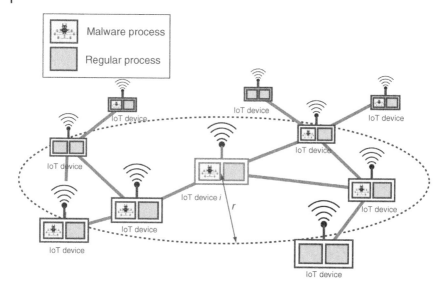

Figure 8.1 Network model: A typical IoT device, referred to as device *i*, is represented as the center node. Each IoT device executes a regular process (indicated by light gray boxes) and may or may not be running a malware process (indicated by boxes with a bot symbol if infected or a dark gray box otherwise). Devices within the communication range (indicated by the dotted line for device *i*) of each other are assumed to be able to communicate with each other and the communication links are highlighted by lines connecting the devices.

$\mathbb{P}[K = k] = \pi_k = \frac{e^{-\lambda \pi r^2} (\lambda \pi r^2)^k}{k!}$. Furthermore, the average degree of a typical device is $\mathbb{E}[K] = \lambda \pi r^2$. An illustration of the network setup along with the state at a particular time is provided in Figure 8.1. A realization of a random network is shown where each IoT device is shown to be equipped with a wireless interface and executing a regular process and a malware process (if infected). The device connectivity is represented by edges between devices that are within a distance *r* of each other. The malware and the control commands propagate over these wireless links from one device to another. A simultaneously executing patching process restores the devices to an un-compromised state (illustrated by the gray boxes).

In order to demonstrate the practical applicability of the employed PPP network model and the associated degree profile of the devices, we use location data of WiFi access points in New York City (NYC), referred to as LinkNYC [104]. A map of the locations of hotspots is provided in Figure 8.2a. We use the location data of 652 hotspots located in Midtown Manhattan and surrounding neighborhoods. Assuming the wireless IoT devices are deployed at the locations of LinkNYC hotspots with a communication range of 140 m, the connectivity profile of a typical devices will almost be Poisson distributed.[1] The empirical degree distribution along with the maximum likelihood estimated Poisson degree is shown in Figure 8.2. Some distortion is observed due to the physical limitation on the hotspots to be confined to the Manhattan grid lines.

We assume that the network is uncoordinated and the devices communicate with each other using (ALOHA) [1] as the MAC protocol. In other words, the devices do not coordinate with each other in making transmission decisions.[2] A significant amount of literature is available on capturing the effects of interference, characterizing the probability of transmission success, and evaluating transmission capacity in Poisson wireless ad hoc networks [145]. In this chapter, we introduce the probability of transmission success of a typical transmitting device as a parameter $\rho \in [0, 1]$. Precise characterization can be obtained using tools from stochastic geometry [133], such as in [57, 66]; however it is not the main focus of this work.

8.1.2 Threat Model

We assume that a botmaster, i.e. the entity that has authored the malware and subsequently plans to launch an attack, possesses powerful capabilities to exploit loopholes in vulnerable wireless IoT devices to infiltrate them and install malicious software process on them. We assume that a proportion $p \in [0, 1]$ of the network is vulnerable to being compromised or infiltrated by the malware if the malware has been successfully transmitted over the wireless interface.[3] In other words, $1/p$ can be considered to be the average number of successful transmission attempts required to infiltrate a neighboring device.

The bots use a fraction of the communication resources of the host device to infiltrate nearby devices and to share control commands. The transmission rate of packets to break into other devices is referred to as *malware spreading rate* and denoted by $\gamma_b \geq 0$ in units of packets per second. Similarly, the transmission rate of packets contributing toward the dissemination of control commands is referred to as *control command propagation rate* and denoted by $\gamma_c \geq 0$. Note that the sum of γ_b and γ_c must be sufficiently small in order to maintain stealthy operation of the botnet.

In summary, the botnet threat in the wireless IoT networks is twofold. Firstly, the malware may spread from one device to another in its proximity using the

---

1 Note that the LinkNYC data has been used as an example to demonstrate the idea of wireless device reachability in large scale public/privately deployed IoT devices in the future.

2 Note that the subsequently proposed framework is not restrictive to a particular MAC protocol. Other MAC protocols such as the carrier sense multiple access (CSMA) can also be used; however, the mean-field dynamics may not directly apply.

3 Vulnerability to be compromised can emanate from events such as using default passwords for access control, using an older version of the firmware, etc.

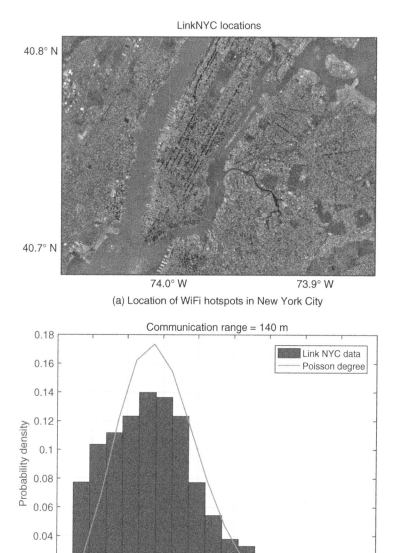

(a) Location of WiFi hotspots in New York City

(b)

Figure 8.2 Analyzing potential connectivity of WiFi hotspots in NYC. Source: (a) LinkNYC.

wireless interface. Secondly, the infected devices referred to as bots share control commands using the same wireless medium to coordinate and plan for launching a network-wide attack. However, as soon as a particular device is patched, the malicious process running on the device is terminated and it gets rid of both the malware as well as information about the control commands. After being patched, the device becomes vulnerable to infection again in the future.[4]

8.2 Propagation Modeling and Analysis

In this section, we provide a systematic approach to model the propagation of malware and formation of a botnet in wireless IoT networks. The proposed model is formally described using the dynamics of population processes and the analysis of equilibrium is presented. Finally, a network defense problem is formulated and a polynomial time algorithm is proposed to obtain the optimal device patching strategy mitigating the formation of a botnet and associated risk of network-wide attack.

8.2.1 Modeling of Malware and Information Evolution

In a large scale wireless IoT network, a typical device may be either un-compromised or infiltrated by malware, thus referred to as a *bot*. Furthermore, devices that are bots may or may not have received control commands. Those that have received control commands may have discarded them due being stale or outdated. Note that since the devices may go from one state to the other based on their communication interactions within their neighborhood, it is appropriate to categorize the devices according to their connectivity or degree.[5] This allows us to use the degree based mean field approach to study the spread of malware and their communication [109]. The possible system states of the population of degree k devices, i.e. devices that are capable of communicating with k other devices, can then be classified as follows:

- $\tilde{B}_k$ – The proportion of degree k devices that are un-compromised.
- $B\tilde{I}_k$ – The proportion of degree k devices that are bots but uninformed about control commands.

---

4 In practice, the device vulnerability for future infection may reduce after getting patched; however, there always exists a certain minimum vulnerability level of the devices. Moreover, the botmaster may also update its strategies to render the devices vulnerable again.

5 This implies that devices with similar connectivity profile will have similar behavior in terms of botnet formation.

- BI_k – The proportion of degree k devices that are bots and are also informed with control commands.

Once, the states are defined, we can study the transitions between each of these states. At any given time an un-compromised device may become an un-informed bot at a rate that is proportional to its degree k and the average probability that it is connected to a bot device, denoted by σ_1. Similarly, an un-informed bot may become an informed bot at a rate that is proportional to its degree k and the average probability that it is connected to an informed bot, denoted by σ_2. On the other hand, an informed bot may discard the control commands at a constant rate β to return to an un-informed state to maintain recency of control information. Finally, if the bots are patched, they return to an un-compromised state. We use a degree based patching rate μ_k inspired from the non-uniform transmission model proposed in [37]. This completes all the transitions between the possible system states.

8.2.2 State Space Representation and Dynamics

In this section, we formally express the dynamics of the system using the developed state space. The state space representation and associated transitions described in Section 8.2.1 are illustrated by the state diagram shown in Figure 8.3. Using the figure and leveraging concepts from the theory of population processes [70], the state evolution can be mathematically described by the following dynamical system of equations:

$$\frac{d\tilde{B}_k(t)}{dt} = \mu_k(B\tilde{I}_k(t) + BI_k(t)) - k\sigma_1\tilde{B}_k(t),$$

$$= \mu_k(1 - \tilde{B}_k(t)) - k\sigma_1\tilde{B}_k(t), \tag{8.1}$$

$$\frac{dB\tilde{I}_k(t)}{dt} = -(\mu_k + k\sigma_2)B\tilde{I}_k(t) + k\sigma_1\tilde{B}_k(t) + \beta BI_k(t), \tag{8.2}$$

$$\frac{dBI_k(t)}{dt} = -(\mu_k + \beta)BI_k(t) + k\sigma_2 B\tilde{I}_k(t). \tag{8.3}$$

Note that (8.1) captures the birth and death processes of un-compromised devices. In other words, it implies that at time t, the population proportion of un-compromised degree k devices is increasing with a rate that is proportional to the patching rate and the population proportion of bot devices. However, at the same time, it is also decreasing at a rate that is proportional to the degree k and the expected rate of interacting with a bot device. Similarly, we can interpret the remaining dynamical equations for un-informed bot and informed bot populations. Since the states represent the population proportions, we can use the closure relationship, i.e. $\tilde{B}_k(t) + B\tilde{I}_k(t) + BI_k(t) = 1, \forall t \geq 0$, to reduce

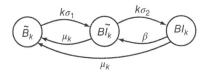

Figure 8.3 State evolution diagram for a typical device. Un-compromised devices of degree k, represented by $(\tilde{B}_k)$, may become infected with malware to become un-informed bot devices $(\tilde{Bl}_k)$, which can further become informed bots (Bl_k). The informed devices discard information at a rate β to again become un-informed. A patching process brings both un-informed and informed bots to an un-compromised state.

Eqs. (8.1)–(8.3) to the following independent dynamical system of equations:

$$\frac{d\tilde{B}_k(t)}{dt} = \mu_k - (\mu_k + k\sigma_1)\tilde{B}_k(t), \tag{8.4}$$

$$\frac{dBl_k(t)}{dt} = k\sigma_2 - (\mu_k + \beta + k\sigma_2)Bl_k(t) - k\sigma_2\tilde{B}_k(t). \tag{8.5}$$

Note that the average probability for a degree k device to be connected to a bot device, σ_1, is directly proportional to the probability of transmission success, the vulnerability of the devices, the malware spreading rate, and the probability of being connected to a bot device. Similarly, the average probability for a degree k device to be connected to an informed bot, σ_2, is directly proportional to the probability of transmission success, the control command propagation rate, and the probability of being connected to an informed bot device. These can be, respectively, expressed as follows:[6]

$$\sigma_1 = \rho\gamma_b p(1 - \theta_{\tilde{B}}), \tag{8.6}$$

$$\sigma_2 = \rho\gamma_c \theta_{BI}, \tag{8.7}$$

where $\theta_{\tilde{B}}$ is the probability that a particular link of a degree k device points to an un-compromised device, and θ_{BI} is the probability that a particular link of a degree k device points to an informed bot device. These probabilities can be evaluated as $\theta_{\tilde{B}} = \sum_{k'}\mathbb{P}(k'|k)\tilde{B}_{k'}(t)$ and $\theta_{BI} = \sum_{k'}\mathbb{P}(k'|k)BI_{k}(t)$. However, for networks with uncorrelated degrees, these probabilities can be further expressed as follows:

$$\theta_{\tilde{B}} = \sum_{k'}\frac{k'P(k')}{\mathbb{E}[K]}\tilde{B}_{k'}(t), \tag{8.8}$$

$$\theta_{BI} = \sum_{k'}\frac{k'P(k')}{\mathbb{E}[K]}BI_{k'}(t). \tag{8.9}$$

---

6 The event of a device being vulnerable to malware infection and the successful reception of wireless signals are independent. Hence, the probabilities can be directly multiplied.

Note that the dynamical system of equations in Eqs. (8.4) and (8.5) describe the time evolution of the respective populations of un-compromised and informed bot devices in the network over time. In order to determine the eventual levels of each type of population in the network, we need to evaluate the equilibrium of the dynamical system. In the subsequent, subsections we focus on analyzing the equilibrium populations of degree k devices.

8.2.3 Analysis of Equilibrium State

At the equilibrium state, $\frac{d\tilde{B}_k(t)}{dt} = 0$ and $\frac{dBI_k(t)}{dt} = 0$. Therefore, the equilibrium population of degree k un-compromised devices, $\tilde{B}_k^*$, and of informed bot devices, BI_k^*, can be expressed as follows:

$$\tilde{B}_k^*(\mu_k) = \frac{\mu_k}{\mu_k + k\sigma_1(\theta_{\tilde{B}}^*)}, \tag{8.10}$$

$$BI_k^*(\mu_k) = \frac{k^2\sigma_1(\theta_{\tilde{B}}^*)\sigma_2(\theta_{BI}^*)}{(\mu_k + k\sigma_1(\theta_{\tilde{B}}^*))(\beta + \mu_k + k\sigma_2(\theta_{BI}^*))}, \tag{8.11}$$

with $\theta_{\tilde{B}}^*$ and θ_{BI}^* denoting the respective probabilities at equilibrium. Note that Eqs. (8.8) and (8.9) express $\tilde{B}_k^*$ and BI_k^* in terms of $\theta_{\tilde{B}}^*$ and θ_{BI}^*. However, Eqs. (8.10) and (8.11) can be used to express $\theta_{\tilde{B}}^*$ and θ_{BI}^* in terms of $\tilde{B}_k^*$ and BI_k^*. Therefore, it presents a self-consistent system of equations, which needs to be solved in order to obtain the equilibrium state. An exact solution to the system is analytically challenging. However, an approximate characterization[7] of the probabilities $\theta_{\tilde{B}}$ and θ_{BI} at equilibrium is provided by the following lemma.

Lemma 8.1 *In a PPP distributed wireless network with D2D communication, the probability of a particular link of a degree k device pointing to an un-compromised and to an informed bot device respectively at equilibrium can be approximately expressed as follows:*

$$\theta_{\tilde{B}}^* \approx \min\left(\frac{\mu_k}{\rho\gamma_b p\mathbb{E}[K]}, 1\right), \tag{8.12}$$

$$\theta_{BI}^* \approx \max\left(1 - \frac{\mu_k\gamma_c + \rho\gamma_b(\beta + \mu_k)}{\mathbb{E}[K]\rho p\gamma_b\gamma_c}, 0\right). \tag{8.13}$$

---

7 Note that these results are based on first order approximation of the first moment of a function of a random variable. Although higher order approximations would lead to tighter approximations, it makes the solution analytically complicated precluding subsequent analysis and optimization.

Proof: By substituting (8.10) into (8.8), we arrive at the following equation that needs to be solved for $\theta_{\tilde{B}}$:

$$\theta_{\tilde{B}} = \sum_{k'} \frac{k'P(k')}{\mathbb{E}[K]} \left(\frac{\mu_{k'}}{\mu_{k'} + k'\sigma_1(\theta_{\tilde{B}})} \right),$$

$$= \sum_{k'} \frac{k'P(k')}{\mathbb{E}[K]} \left(\frac{\mu_{k'}}{\mu_{k'} + k'\rho\gamma_b p(1 - \theta_{\tilde{B}})} \right). \tag{8.14}$$

The optimal $\theta_{\tilde{B}}$ is referred to as $\theta_{\tilde{B}}^*$. The first step is to make use of the degree independence in a homogeneous PPP network to write (8.14) as follows:

$$\theta_{\tilde{B}}^* = \mathbb{E} \left[\frac{\mu_k}{\mu_k + k\rho\gamma_b p(1 - \theta_{\tilde{B}}^*)} \right]. \tag{8.15}$$

Due to the complex form of $\mathbb{P}(K = k)$, a tractable closed form for $\mathbb{E} \left[\frac{\mu_k}{\mu_k + k\rho\gamma_b p(1 - \theta_{\tilde{B}}^*)} \right]$ cannot be easily obtained. Using Taylor expansions for the moments of functions of random variables, the expectation of a function $g(\cdot)$ can be expressed as $\mathbb{E}[g(K)] \approx g(\mathbb{E}[K]) + \frac{g''(\mathbb{E}[K])}{2}\sigma_K^2$, where σ_K is the variance of the degree. However, using a second order approximation results in loss of tractable solution for (8.15). Therefore, we resort to the first order approximation for simplicity, which results in (8.15) being expressed as follows:

$$\theta_{\tilde{B}}^* \approx \frac{\mu_k}{\mu_k + \mathbb{E}[K]\rho\gamma_b p(1 - \theta_{\tilde{B}}^*)}. \tag{8.16}$$

It can be solved for $\theta_{\tilde{B}}^*$ to lead to the following:

$$\theta_{\tilde{B}}^* \approx \frac{\mu_k}{\rho\gamma_b p\mathbb{E}[K]}. \tag{8.17}$$

Note that since $\mu_k \geq 0$ is not bounded from above, so $\theta_{\tilde{B}}^*$ may become higher than unity which is not possible since it represents a probability. Therefore, we restrict it from above by unity, thus proving the first part of the lemma. Using a similar methodology, substituting (8.11) into (8.9) leads to the following expression for θ_{BI}^*:

$$\theta_{BI}^* = \mathbb{E} \left[\frac{k^2\sigma_1(\theta_{\tilde{B}})\sigma_2(\theta_{BI}^*)}{(\mu_k + k\sigma_1(\theta_{\tilde{B}}))(\beta + \mu_k + k\sigma_2(\theta_{BI}^*))} \right],$$

$$= \mathbb{E} \left[\frac{k^2\sigma_1(\theta_{\tilde{B}})\rho\gamma_c\theta_{BI}^*}{(\mu_k + k\sigma_1(\theta_{\tilde{B}}))(\beta + \mu_k + k\rho\gamma_c\theta_{BI}^*)} \right]. \tag{8.18}$$

Again, using the first order approximation of the function inside the expectation, we arrive at solving the following equation:

$$\theta_{BI}^* \approx \frac{(\mathbb{E}[K])^2\sigma_1(\theta_{\tilde{B}})\sigma_2(\theta_{BI}^*)}{(\mu_k + \mathbb{E}[K]\sigma_1(\theta_{\tilde{B}}))(\beta + \mu_k + \mathbb{E}[K]\sigma_2(\theta_{BI}^*))}, \tag{8.19}$$

Solving this for θ_{BI}^*, after some algebraic manipulations, leads to the following result:

$$\theta_{BI}^* \approx 1 - \frac{\mu_k \gamma_c + \rho \gamma_b (\beta + \mu_k)}{k \rho p \gamma_b \gamma_c}. \tag{8.20}$$

Since μ_k represents a probability, it needs to be non-negative. Hence, θ_{BI}^* needs to be restricted at 0 from below, leading to the result provided in Lemma 8.1. In Figure 8.4, we plot the results obtained from the first order and second order approximations of the probabilities $\theta_{\tilde{B}}^*$ and θ_{BI}^* against the patching rates. It is observed that the gap between the approximations increases as the patching rate gets higher. Furthermore, the approximations for $\theta_{\tilde{B}}^*$ are relatively much closer as compared with the ones for θ_{BI}^*. Therefore, despite some loss in accuracy, it is still reasonable to use the first order approximations due to the powerful analytical tractability that facilitates further analysis and decision-making. □

These approximations present a lower bound on the actual probabilities. The loss in accuracy for the sake of analytical tractability is discussed in the proof of Lemma 8.1. Note that Lemma 8.1 presents an intuitive result where the probability of being connected to an un-compromised device, $\theta_{\tilde{B}}$, is directly proportional to the patching rate and inversely related to the expected degree, vulnerability, malware spreading rate, and transmission success probability. Similar explanation can be derived for θ_{BI}. A direct corollary of the result presented in Lemma 8.1, that plays an important role in the optimal patching decisions is provided below:

Corollary 8.1 *For a PPP deployed wireless IoT network being infiltrated by a botnet with malware spreading at a rate γ_b and control commands propagating at a rate γ_c, the upper bound on the required patching rate for a device to have an impact on the equilibrium populations is given by*

$$\mu_k \leq \rho \gamma_b p \mathbb{E}[K], \quad \forall k \geq 1, \tag{8.21}$$

Proof: From (8.10), we deduce that in order for $\theta_{\tilde{B}}^*$ to assume a nontrivial value, $\frac{\mu_k}{\rho \gamma_b p \mathbb{E}[K]}$ must be smaller than unity. This implies that $\mu_k \leq \rho \gamma_b p \mathbb{E}[K]$. □

This is significant since it provides an estimate of the maximum patching frequency that can be used by the network defender on a degree k device to have an impact on the equilibrium proportions of the devices. In other words, it presents the fundamental limits of the patching rate, since using a higher patching rate will lead to a completely bot-free population at equilibrium. Similarly, an auxiliary result emanating from (8.13) is expressed in the following corollary.

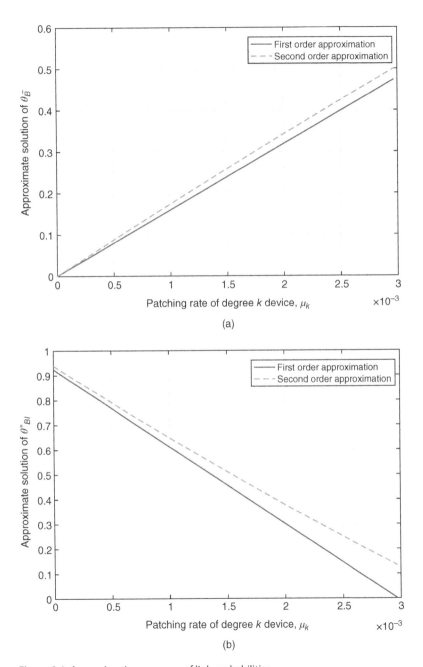

Figure 8.4 Approximation accuracy of link probabilities.

Corollary 8.2 *The maximum information refresh rate, β, that can be selected by a bot device to have non-zero informed bot population at equilibrium can be expressed as follows:*

$$\beta < p\gamma_c \mathbb{E}[K]. \tag{8.22}$$

Proof: From (8.11), we deduce that $\frac{\mu_k \gamma_c + p\gamma_b(\beta + \mu_k)}{\mathbb{E}[K]\rho p \gamma_b \gamma_c} \leq 1$ in order for θ_{BI}^* to assume a non-trivial value. It results in the condition $\mu_k \leq \frac{p\gamma_b\gamma_c p\mathbb{E}[K] - p\gamma_b\beta}{\gamma_c + p\gamma_b}$ with an implicit condition $\beta < p\gamma_c\mathbb{E}[K]$ for it to be meaningful. It is formally expressed as Corollary 8.2. However, the upper bound obtained from (8.10) is higher, thus becoming the effective upper bound. Therefore, any μ_k higher than the upper bound is futile in having an impact on the equilibrium state of the devices. In other words, patching devices at a higher rate than the upper bound only affects the regular network operation without having any impact on botnet formation. □

Although the results presented in Lemma 8.1 are useful, the presence of the minimum and maximum functions presents a challenge in leveraging them for optimization purposes. To circumvent this challenge, we propose to use the Log-Sum-Exponential (LSE) function[8] to provide a smooth and continuously differentiable approximation of these expressions. It results in the following:

$$\theta_{\tilde{B}}^* \approx -\frac{1}{\eta} \ln \left(e^{-\eta} + e^{-\eta\left(\frac{\mu_k}{p\gamma_b p\mathbb{E}[K]} \right)} \right), \tag{8.23}$$

$$\theta_{BI}^* \approx \frac{1}{\eta} \ln \left(1 + e^{\eta\left(1 - \frac{\mu_k\gamma_c + p\gamma_b(\beta + \mu_k)}{\mathbb{E}[K]\rho p\gamma_b\gamma_c} \right)} \right), \tag{8.24}$$

where η is a sufficiently large constant chosen for accuracy of the *soft-minimum* and *soft-maximum* functions. Note that the LSE relaxation in Eqs. (8.23) and (8.24) may slightly affect the upper bound on the patching rate expressed in Corollary 8.1 and the upper bound on the possible information refresh rate expressed in Corollary 8.2. However, the inaccuracy diminishes with the selection of large η.

$$\tilde{B}_k^*(\mu_k) \approx \frac{\mu_k}{\mu_k + k\rho\gamma_b p \left(1 + \frac{1}{\eta} \ln \left(e^{-\eta} + e^{-\eta\left(\frac{\mu_k}{p\gamma_b p\mathbb{E}[K]} \right)} \right) \right)}, \tag{8.25}$$

---

8 The function max(x, y) can be approximated by $\frac{1}{\eta} \ln(e^{\eta x} + e^{\eta y})$ and min(x, y) can be approximated by $-\frac{1}{\eta} \ln(e^{-\eta x} + e^{-\eta y})$ provided that η is sufficiently large.

$$BI_k^*(\mu_k) \approx \frac{k^2\rho^2\gamma_b\gamma_c p\left(1 + \frac{1}{\eta}\ln\left(e^{-\eta} + e^{-\eta\left(\frac{\mu_k}{\rho\gamma_b p E[K]}\right)}\right)\right)}{\left(\mu_k + k\rho\gamma_b p\left(1 + \frac{1}{\eta}\ln\left(e^{-\eta} + e^{-\eta\left(\frac{\mu_k}{\rho\gamma_b p E[K]}\right)}\right)\right)\right)}$$

$$\times \frac{\frac{1}{\eta}\ln\left(1 + e^{\eta\left(1 - \frac{\mu_k\gamma_c + \rho\gamma_b(\beta+\mu_k)}{E[K]\rho\gamma_b\gamma_c}\right)}\right)}{\left(\beta + \mu_k + k\rho\gamma_c + \frac{1}{\eta}\ln\left(1 + e^{\eta\left(1 - \frac{\mu_k\gamma_c + \rho\gamma_b(\beta+\mu_k)}{E[K]\rho\gamma_b\gamma_c}\right)}\right)\right)}. \tag{8.26}$$

Finally, using the results of Lemma 8.1 and the subsequent LSE relaxation, the equilibrium populations of devices that are un-compromised and devices that are informed bots are expressed by the following theorem:

Theorem 8.1 *At equilibrium, the proportion of degree k devices in the network that are un-compromised (not infected with malware), i.e. $\tilde{B}_k^*$ and those that are bots and informed by control commands, i.e. BI_k^* can be approximately expressed by Eqs. (8.25) and (8.26) respectively.*

Proof: Substitution of (8.23) into (8.10) and (8.24) into (8.11) leads to this result. □

In Section 8.3, we make use of the developed analytical model and the approximate results to formulate the network defense problem and subsequently discuss the methodology for solving it.

8.3 Patching Mechanism for Network Defense

The goal of the network defender is to set up a patching schedule for each network device based on its connectivity in order to prevent the formation of a large scale botnet. The patching rate must take into account the disruption caused to regular operation due to the strategies employed, e.g. firmware upgrade or power cycling, which can be in terms of the downtime of devices. The cost incurred on the operation of a network device due to patching activity is assumed to be a smooth, convex, and increasing function of the patching rate μ_k, represented by $\phi_k : \mathbb{R}^+ \to \mathbb{R}^+, \forall k \geq 1$. The risk of a botnet formation can be measured in terms of the equilibrium population of devices that are bots and the devices that are receiving control commands assuming knowledge of the transmission rates. Accordingly, targets for the minimum expected proportion of network that is un-compromised and the maximum tolerable proportion of the network that is an informed bot, denoted by $\tau_{\tilde{B}} \in [0,1]$ and $\tau_{BI} \in [0,1]$

respectively, can be set. The network defender's problem can then be formulated as follows:

$$\underset{\mu_k, k \geq 1}{\text{minimize}} \quad \sum_{k=1}^{\infty} \phi_k(\mu_k)\pi_k, \tag{8.27}$$

$$\text{subject to} \quad \sum_{k=1}^{\infty} \tilde{B}_k^*(\mu_k)\pi_k \geq \tau_{\tilde{B}}, \tag{8.28}$$

$$\sum_{k=1}^{\infty} BI_k^*(\mu_k)\pi_k \leq \tau_{BI}. \tag{8.29}$$

The objective represents the total expected cost of patching devices at a rate $\mu_k, \forall k$, while the constraints imply that the average proportion of un-compromised devices in the network must be higher than $\tau_{\tilde{B}}$ and the average proportion of informed bot devices in the network must be smaller than τ_{BI}. Note that the constraints are coupled with the objective, which makes the primal problem challenging to solve. Furthermore, despite the fact that the objective is convex, both the constraints may be non-convex in the decision vector since some terms inside the summation are concave while others are convex. This is formally stated in the following lemma.

Lemma 8.2 *The equilibrium proportion of un-compromised devices, $\tilde{B}_k^*$, is concave in μ_k for $k < \mathbb{E}[\mathbb{K}]$ and convex otherwise. Similarly, there is a change in curvature of the equilibrium proportion of informed bot devices, BI_k, from convex to concave with increasing device degree k.*

Proof: We can observe that $\dfrac{d\tilde{B}_k^*}{d\mu_k} = \dfrac{k\sigma_1 - \mu_k k\sigma_1'}{(\mu_k + k\sigma_1)^2}$ and $\dfrac{d^2\tilde{B}_k^*}{d\mu_k^2} =$
$\dfrac{(\mu_k + k\sigma_1)((\mu_k + k\sigma_1)(-\mu_k k\sigma_1'') - 2(1 + k\sigma_1')(k\sigma_1 - \mu_k k\sigma_1'))}{(\mu_k + k\sigma_1)^3}$, where $\sigma_1' = \dfrac{d\sigma_1(\mu_k)}{d\mu_k} = -\dfrac{1}{\mathbb{E}[K]}$ and $\sigma_1'' = \dfrac{d^2\sigma_1(\mu_k)}{d\mu_k^2} = 0$. The denominator of $\dfrac{d^2\tilde{B}_k^*}{d\mu_k^2}$ is always positive and the numerator evaluates to $-2(\mu_k + k\sigma_1)\left(1 - \dfrac{k}{\mathbb{E}[K]}\right)\left(k\sigma_1 + \dfrac{k\mu_k}{\mathbb{E}[K]}\right)$. Therefore, it is clear that $\dfrac{d^2\tilde{B}_k^*}{d\mu_k^2} < 0$ if $k < \mathbb{E}[K]$ and vice versa. Therefore, we can conclude that $\tilde{B}_k$ evaluated at equilibrium is concave for $k < \mathbb{E}[K]$ and convex otherwise. Similarly, for BI_k, it can be shown that $\dfrac{d^2 BI_k^*}{d\mu_k^2}$ experiences a change in sign with k, which is hard to characterize analytically but the change point can be proved to be different than $\mathbb{E}[K]$. In order to demonstrate the change in curvature of the equilibrium populations, we plot the respective equilibrium populations of un-compromised devices and informed bots in Figure 8.5 for different values of k. Note that with an increasing patching rate, the un-compromised device population increases until it reaches 1 ($-\tilde{B}_k$ is plotted in Figure 8.5, which is

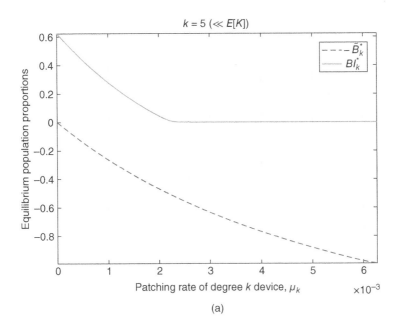

(a)

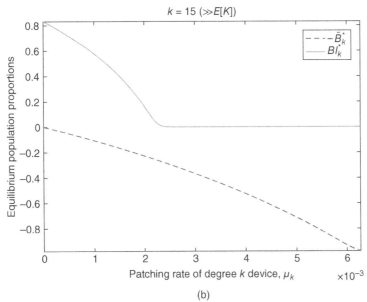

(b)

Figure 8.5 Curvature analysis of equilibrium population processes for different degree devices. (a) Equilibrium populations against patching rate for small degree devices. (b) Equilibrium populations against patching rate for large degree devices.

decreasing to -1). However, on the other hand, the equilibrium population of informed bot devices decreases until it reaches 0. Furthermore, the informed bot device population diminishes completely with a much smaller patching rate that is required to make the network completely un-compromised. These equilibrium populations have been plotted with mean device degree $\mathbb{E}[K] = 9.4$ and it can be observed that the curvature of the constraints is different if the degree is small, i.e. $k = 5$, than when it is large, i.e. $k = 15$. $\square$

Another important observation is that the constraints are linked in terms of the patching rates. A set of patching rates may completely satisfy one of the constraints but not the other. Therefore, it is important to investigate the conditions under which the constraints are active, particularly because there exists a limiting rate at which the constraints saturate. The following lemma presents an important condition relating the target thresholds that determines the status of the constraints.

Lemma 8.3 *The constraint on the average equilibrium population of informed bots, expressed in (8.29), is always satisfied for any $\tau_{BI} \in [0, 1]$ if the target on the average equilibrium population of un-compromised devices is set as follows:*

$$\tau_{\tilde{B}} \geq \frac{\mathbb{E}[K]p\gamma_c - \beta}{\mathbb{E}[K]p(\rho\gamma_b + \gamma_c)} \tag{8.30}$$

Proof: From proof of Lemma 8.1, it can be concluded that $\hat{\mu}_k = \frac{\rho\gamma_b\gamma_c p\mathbb{E}[K] - \rho\gamma_b\beta}{\gamma_c + \rho\gamma_b}$ can completely eradicate equilibrium population of informed bots of degree k. However, at this patching rate, the population of un-compromised devices can be obtained as $\tilde{B}_k^*(\hat{\mu}_k) = \frac{\mathbb{E}[K](p\gamma_c\mathbb{E}[K] - \beta)}{k(\beta + \mathbb{E}[K]p\rho\gamma_b) + \mathbb{E}[K](\mathbb{E}[K]p\gamma_c - \beta)}$. Since $\tilde{B}_k^*(\hat{\mu}_k)$ is a convex function of k, $\sum_{k=1}^{\infty} \tilde{B}_k^*(\hat{\mu}_k) = \mathbb{E}[\tilde{B}_k^*(\hat{\mu}_k)] \geq \tilde{B}_{\mathbb{E}[K]}^*(\hat{\mu}_k)$ (using Jensen's inequality [27]). It results in $\mathbb{E}[\tilde{B}_k^*(\hat{\mu}_k)] \geq \frac{\mathbb{E}[K]p\gamma_c - \beta}{\mathbb{E}[K]p(\rho\gamma_b + \gamma_c)}$. Knowing that $\tilde{B}_k^*$ is an increasing function of μ_k, we can deduce that if $\tau_{\tilde{B}} \geq \frac{\mathbb{E}[K]p\gamma_c - \beta}{\mathbb{E}[K]p(\rho\gamma_b + \gamma_c)}$, then it requires a patching rate higher than $\hat{\mu}_k$. This implies that BI_k^* will be zero at the optimal patching rate. Hence, the constraint (8.29) will always be satisfied if $\tau_{\tilde{B}}$ is sufficiently high, and therefore, we can effectively remove it from the optimization problem. This phenomenon can also be observed from Figure 8.5 where the equilibrium population of informed bots diminishes to zero much earlier than the equilibrium proportion of un-compromised devices. $\square$

Therefore, if the condition presented in Lemma 8.3 is satisfied, we can effectively ignore the constraint (8.29) from the optimization problem and proceed with only (8.28). This is extremely important since otherwise, the solution to the optimization problem may be difficult as one of the constraints saturates and is no longer monotonously increasing or decreasing. However, evaluating the condition *a priori*, we can circumvent this difficulty and effectively solve the optimization problem. However, there are several additional challenges. First, since the network is random, there is no upper bound on the maximum possible degree of a device, which makes the optimization problem intractable due to an infinite number of optimization variables. However, due to the structure of the network,[9] it is increasingly rare for a device to have larger degrees. Therefore, we note that there exists a sufficiently large $k = k_{\max}$ such that $\mathbb{P}[K > k_{\max}] \leq \varepsilon$, where ε is arbitrarily small. This allows us to convert the optimization problem into one with finite number of optimization variables referred to as $\mu = [\mu_1, \mu_2, \ldots, \mu_{k_{\max}}]^T$. Therefore, the problem can then be expressed as follows:

$$\underset{\mu}{\text{minimize}} \sum_{k=1}^{k_{\max}} \phi_k(\mu_k)\pi_k + \underbrace{\sum_{k=k_{\max}}^{\infty} \phi_k(\mu_k)\pi_k}_{\varepsilon_1}$$

$$\text{subject to } \tau_{\tilde{B}} - \sum_{k=1}^{k_{\max}} \tilde{B}_k^*(\mu_k)\pi_k - \underbrace{\sum_{k=k_{\max}}^{\infty} \tilde{B}_k^*(\mu_k)\pi_k}_{\varepsilon_2} \leq 0, \tag{8.31}$$

$$\sum_{k=1}^{k_{\max}} BI_k^*(\mu_k)\pi_k + \underbrace{\sum_{k=k_{\max}}^{\infty} BI_k^*(\mu_k)\pi_k}_{\varepsilon_3} - \tau_{BI} \leq 0.$$

Since the Poisson density decays faster than the exponential rate for large degree values, the terms labeled as $\varepsilon_1, \varepsilon_2$, and ε_3 can be made arbitrarily small for sufficiently large $k_{\max}$. Hence, effectively, these terms can be removed and the problem can be converted into a finite optimization problem. Since the primal problem may be non-convex, we resort to solving the dual optimization problem [155]. Note, however, that the duality gap in this problem setting is zero and hence solving the dual problem is equivalent to solving the primal problem. We, therefore, relax the original problem by forming the Lagrangian as follows:

$$\mathcal{L}(\mu, \zeta, \xi) = \sum_{k=1}^{k_{\max}} \phi_k(\mu_k)\pi_k - \zeta\left(\sum_{k=1}^{k_{\max}} \tilde{B}_k^*(\mu_k)\pi_k - \tau_{\tilde{B}}\right) - \xi\left(\tau_{BI} - \sum_{k=1}^{k_{\max}} BI_k^*(\mu_k)\pi_k\right),$$

$$= \sum_{k=1}^{k_{\max}} (\phi_k(\mu_k)\pi_k - \zeta\tilde{B}_k^*(\mu_k)\pi_k + \xi BI_k^*(\mu_k)\pi_k) + \zeta\tau_{\tilde{B}} - \xi\tau_{BI}. \tag{8.32}$$

---

9 In a PPP network, the probability of having a large number of neighbors decreases faster than the exponential decay rate for sufficiently large degrees.

where ζ and ξ are the Lagrange multipliers, which are dual feasible if $\zeta \geq 0$ and $\xi \geq 0$. The Lagrange dual function can be written as follows:

$$g(\zeta, \xi) = \min_{\mu \geq 0} \sum_{k=1}^{k_{max}} (\phi_k(\mu_k)\pi_k - \zeta \tilde{B}_k^*(\mu_k)\pi_k + \xi BI_k^*(\mu_k)\pi_k) + \zeta \tau_{\tilde{B}} - \xi \tau_{BI},$$

$$= \sum_{k=1}^{k_{max}} (\min_{\mu_k \geq 0} \{\phi_k(\mu_k)\pi_k - \zeta \tilde{B}_k^*(\mu_k)\pi_k + \xi BI_k^*(\mu_k)\pi_k\}) + \zeta \tau_{\tilde{B}} - \xi \tau_{BI}.$$

Note that due to the structure of the Lagrangian, the optimization problem in the dual function decouples in the optimization variables, which makes the complexity of evaluating $g(\zeta, \xi)$ linear in k_{max} [108]. For a given pair of Lagrange multipliers, the optimal patching rates μ^* can be written as follows:

$$\mu_k^* = \arg \min_{\mu_k \geq 0} \{\phi_k(\mu_k)\pi_k - \zeta \tilde{B}_k^*(\mu_k)\pi_k + \xi BI_k^*(\mu_k)\pi_k\}. \tag{8.33}$$

Note that if both $\tilde{B}_k^*$ and BI_k^* are not monotonous in μ_k, it may not be possible to obtain a globally optimal solution for μ_k in (8.33). However, fortunately using Lemma 8.3, we can determine if one of the functions will saturate or not at the optimal μ_k based on the target thresholds set by the defender. If the condition in Lemma 8.3 is satisfied, we can ignore the term containing BI_k^* in (8.33) and proceed with the optimization.[10] Finally, the dual optimization problem can be written as follows:

$$\underset{\zeta \geq 0, \xi \geq 0}{\text{maximize}} \ g(\zeta, \xi). \tag{8.34}$$

Since $g(\zeta, \xi)$ is a concave optimization problem and has a unique maxima, we can employ a gradient based strategy to achieve the optimal result. However, since a closed form of the dual function may not exist, and hence differentiability may not be guaranteed, we can resort to sub-gradient based iterative update methods for the dual variables [146]. The sub-gradients of the dual function, evaluated at the optimal patching rates, can be expressed as follows:

$$\nabla_\zeta g(\zeta, \xi) = \tau_{\tilde{B}} - \sum_{k=1}^{k_{max}} \tilde{B}_k^*(\mu_k^*)\pi_k, \tag{8.35}$$

$$\nabla_\xi g(\zeta, \xi) = \sum_{k=1}^{k_{max}} BI_k^*(\mu_k^*)\pi_k - \tau_{BI}. \tag{8.36}$$

---

10 Removal of the term containing BI_k^* automatically results in the removal of the Lagrange multiplier ξ in the subsequent expressions.

Therefore, the iterative dual update rule based on the sub-gradients can be expressed as follows:

$$\zeta^{(i+1)} = [\zeta^{(i)} - \alpha \nabla_\zeta]^+ = \left[\zeta^{(i)} - \alpha \left(\tau_{\tilde{B}} - \sum_{k=1}^{k_{max}} \tilde{B}_k^*(\mu_k^*) \pi_k \right) \right]^+, \quad i = 0, 1, 2, \ldots,$$

(8.37)

$$\xi^{(i+1)} = [\xi^{(i)} - \alpha \nabla_\xi]^+ = \left[\xi^{(i)} - \alpha \left(\sum_{k=1}^{k_{max}} BI_k^*(\mu_k^*) \pi_k - \tau_{BI} \right) \right]^+, \quad i = 0, 1, 2, \ldots,$$

(8.38)

where α is the step size. The complete procedure for obtaining the optimal patching policy is provided in Algorithm 8.1. We initialize the iteration counter i to zero. Furthermore, we initialize the Lagrange multipliers to an arbitrary positive value and set a sufficiently small step-size α. Based on the condition $\tau_{\tilde{B}} \geq \frac{\mathbb{E}[K]p\gamma_c - \beta}{\mathbb{E}[K]p(\rho\gamma_b + \gamma_c)}$, we exclude or include the term containing BI_k^* and the associated Lagrange multiplier ξ. We then proceed to solve the optimization problem in (8.33) for all possible device degrees. Once the optimal intermediate patching rates have been determined, the dual variables are updated based on the sub-gradient based update rule defined in (8.37) and (8.38). This process is repeated until the dual variables have converged and the corresponding μ_k^*, $\forall k = 1, 2, \ldots, k_{max}$, define the optimal patching rates for each device type. The complete procedure can be shown to have polynomial complexity in the total number of device degrees k_{max}. In Section 8.3.1, we provide numerical studies to illustrate the behavior of the solutions and its sensitivities with respect to different model parameters.

8.3.1 Simulation Results

In this section, we first describe the network setup and system parameters used for numerical studies. Then, we present the results obtained from the solution to the optimization problem and the associated impact of the parameters involved. The parameters selected for the generation of numerical results are for illustrative purposes and can be modified according to the scenario in practical applications.

Consider a random network of wireless IoT devices distributed according to a homogeneous PPP with intensity $\lambda = 300$ device/km^2 and a communication range of $r = 100$ m. On average, a typical IoT device would be able to communicate with $\mathbb{E}[K] = \lambda \pi r^2 = 9.4$, i.e. approximately 9 other devices. We assume that the maximum possible degree in the network is $k_{max} = 25$ for which $\varepsilon = \mathbb{P}(K \geq k_{max})$ is of the order 10^{-6}. Due to interference and fading effects of the wireless channel during communication, we assume a successful transmission probability of $\rho = 0.95$. We assume that a proportion $p = 0.7$ of the

---

Algorithm 8.1 Dual Algorithm to Solve the Optimal Patching Problem

---

Require: Target thresholds, $\tau_{\bar{B}}$ and τ_{BI}.

Require: Iteration $i = 0$, Step-size α, Lagrange multipliers $\zeta^{(i)} > 0, \xi^{(i)} > 0$.

1: **repeat**

2:     **function** DUAL FUNCTION OPTIMIZATION

3:         **if** $\tau_{\bar{B}} \geq \frac{\mathbb{E}[K]p\gamma_c - \beta}{\mathbb{E}[K]p(\rho\gamma_b + \gamma_c)}$ **then**

4:             $\mu_k^{(i)*} \leftarrow \underset{\mu_k \geq 0}{\arg\min}\big\{\phi_k(\mu_k)\pi_k - \zeta\tilde{B}_k^*(\mu_k)\pi_k\big\}, \forall k = 1, \dots, k_{\max}.$

5:         **else**

6:             $\mu_k^{(i)*} \leftarrow \underset{\mu_k \geq 0}{\arg\min}\big\{\phi_k(\mu_k)\pi_k - \zeta\tilde{B}_k^*(\mu_k)\pi_k + \xi BI_k^*(\mu_k)\pi_k\big\}, \forall k = $

   $1, \dots, k_{\max}.$

7:         **end if**

8:     **end function**

9:     **function** DUAL VARIABLE UPDATE

10:         **if** $\tau_{\bar{B}} \geq \frac{\mathbb{E}[K]p\gamma_c - \beta}{\mathbb{E}[K]p(\rho\gamma_b + \gamma_c)}$ **then**

11:         $\zeta^{(i+1)} \leftarrow \left[\zeta^{(i)} - \alpha\left(\tau_{\bar{B}} - \sum_{k=1}^{k_{\max}}\tilde{B}_k^*(\mu_k^*)\pi_k\right)\right]^+.$

12:         $\xi^{(i+1)} \leftarrow \left[\xi^{(i)} - \alpha\left(\sum_{k=1}^{k_{\max}}BI_k^*(\mu_k^*)\pi_k - \tau_{BI}\right)\right]^+.$

13:         **else**

14:         $\zeta^{(i+1)} \leftarrow \left[\zeta^{(i)} - \alpha\left(\tau_{\bar{B}} - \sum_{k=1}^{k_{\max}}\tilde{B}_k^*(\mu_k^*)\pi_k\right)\right]^+.$

15:         **end if**

16:     **end function**

17: **until** convergence of ζ and ξ.

---

network is vulnerable to be infected by malware. The malware introduced by a botmaster is assumed to transmit packets for infiltration in nearby devices at a rate of $\gamma_b = 0.001$ packets per second (or 1 packet every 1000 seconds) and for control commands propagation at a rate of $\gamma_c = 0.001$ packets per second. The information refresh rate of bots is selected as $\beta = 0.002\,\text{s}^{-1}$. Note that this choice of β satisfies the condition provided in Corollary 8.2.

In the theoretical analysis, the scaling constant for LSE relaxation of the minimum function is chosen to be $\eta = 100$ for accuracy. The impact of patching a device of degree k on the operational performance of the network is assumed to be captured by the function $\phi_k(\mu_k) = w_k\mu_k^2$, where the weights are modeled using the following logistic function:

$$w_k = \frac{1}{1 + e^{-a(k-b)}}, \tag{8.39}$$

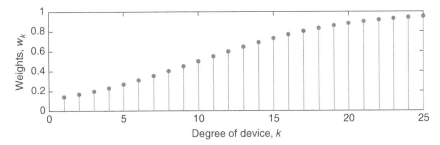

Figure 8.6 Relative impact of unit patching rate of a degree k device on network performance.

and the constants a and b are chosen to be $a = 0.2$ and $b = 10$, respectively. An illustration of the weight function is provided in Figure 8.6. It implies that a unit patching rate on a device of degree k has a higher impact on network operation as k increases. Hence, it is more costly to increase patching rate for higher degree devices.

In Figure 8.7a, we plot the optimal patching rates for a degree k device in the network with varying target of un-compromised device proportion while fixing $\tau_{BI} = 0.2$. The right axis plots the proportion of degree k devices in the network, or equivalently the probability of a typical device having degree k, as a reference for interpreting the results. The dotted line shows the theoretical maximum patching rate that impacts the equilibrium populations as described in Lemma 8.1. It can be observed that for $\tau_{\bar{B}} = 0.6, 0.7$, the optimal patching rates closely follow the proportion of devices due to the monotonously increasing weights w_k. However, for more aggressive targets e.g. $\tau_{\bar{B}} = 0.8, 0.9$, the optimal patching rates saturate for the more probable degrees while increasing patching rates for the less probable ones.

In Figure 8.7b, we plot the optimal patching rates for a degree k device in the network with varying target of informed bot proportion while fixing $\tau_{\bar{B}} = 0.7$. Note that a similar behavior is observed in this case where the optimal patching rates closely follow the network degree profile for less aggressive targets, e.g. $\tau_{BI} = 0.1, 0.2$. However, for more aggressive targets such as $\tau_{BI} = 0.01, 0.05$, a saturation is observed for more probable degree types. However, note that the higher and less probable degree devices are patched more frequently although it causes higher disruption since the targets are otherwise not achievable.

Finally, Figure 8.8a,b illustrate the behavior of the expected total patching cost with varying malware spreading rate and control command spreading rates respectively. It is observed that the expected total patching cost increases at an increasing rate both with increasing malware spreading rate and the target un-compromised device proportion. However, the expected total patching cost increases at a decreasing rate with increasing control command propagation rate. This shows that the defender is more reactive to the malware spreading

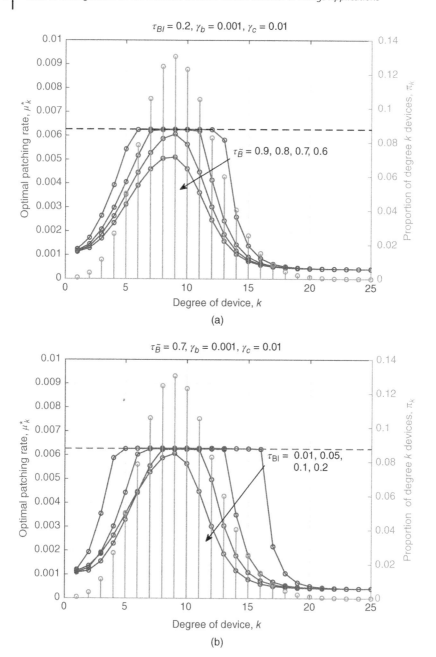

Figure 8.7 Impact of varying un-compromised bot proportion threshold $\tau_{\tilde{B}}$ and informed bot proportion threshold τ_{BI}. The dotted line shows the theoretical upper bound expressed in Corollary 8.1.

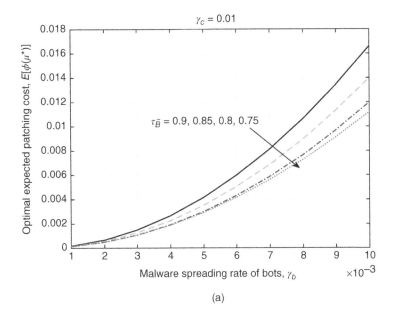

(a)

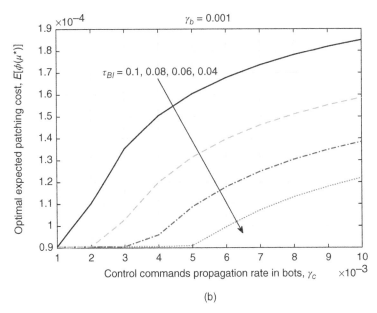

(b)

Figure 8.8 Expected total cost of patching against varying system parameters. (a) Varying malware spreading rate and bot-free population target. (b) Varying control commands propagation rate and informed bot population target.

rate than the control command propagation rate in terms of a botnet formation. With regard to the effect of varying the device vulnerability in the network as well as the probability of transmission success, a similar behavior is observed since changing these parameters in turn alters the effective malware propagation rate and the control command propagation rate.

8.3.2 Simulation and Validation

In this section, we conduct simulation experiments to validate the accuracy of the obtained theoretical results. In the first part, we simulate the considered PPP network. Two different phases are investigated. In the first phase, a malware is introduced at epoch to an arbitrarily selected node and is allowed to propagate to its neighborhood according to the device vulnerabilities, wireless transmission success probability, and malware propagation rates. The malware spreads from one device to another in a D2D fashion until all the network has been compromised. Note that during the initial phase, there is no patching of devices. During the second phase, the optimal patching policy for each device, based on its degree, is applied on the network. This leads to the recovery of bot devices and the proportions of bots in the network is observed over time. The experiment is repeated for different target thresholds for bot-free population, i.e. $\tau_{\bar{B}} = 0.7, 0.8$, and 0.9. Figure 8.9 illustrates a snapshot of the device states in the network after reaching equilibrium. Note that more devices are un-compromised at equilibrium as $\tau_{\bar{B}}$ increases as reflected by Figure 8.9a–c. The time evolution of un-compromised devices for each of the thresholds is recorded in Figure 8.10(a). Notice that the proportion of un-compromised devices increasingly drops from 100% to 0% as the malware is allowed to propagate in the network. However, when the patching process is started in the second phase (i.e. $t = 10^4$), the bot-free population sharply rises until it reaches the target threshold. Although the population keeps fluctuating due to the ongoing dynamical processes but on average the policy is observed to accurately achieve the defined targets.

To further illustrate the usefulness and impact of our proposed methodology and obtained results, we simulate an experiment on the actual LinkNYC hotspot locations data. We assume that IoT devices are placed at each of these locations with a communication range of 140 m. Again, the simulation is carried out in two phases. In the first phase, the malware is allowed to propagate in the network until it has achieved the maximum spread. To ensure complete penetration of the malware in the network, we initially introduce the malware in nodes that have a degree of 2. This allows the propagation of the malware from one device to another over time until it affects most of the nodes during the first phase. Note that this network is not exactly a PPP, the malware spread is not as effective since some nodes may be isolated or clustered together. Similarly, during the second phase (i.e. $t = 2.5 \times 10^4$), the patching process is started until

Figure 8.9 Snapshot of network states at equilibrium in a PPP network.

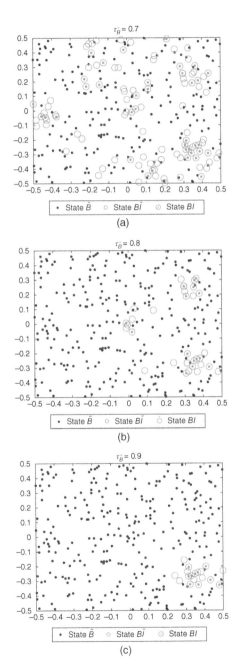

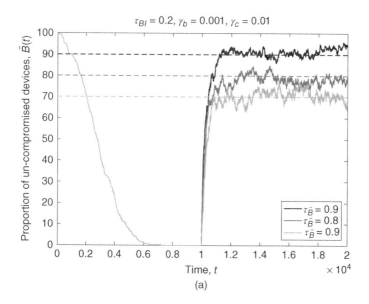

(a)

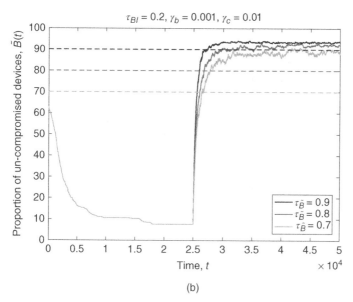

(b)

Figure 8.10 Time evolution of the proportion of un-compromised devices. (a) PPP network and (b) LinkNYC network.

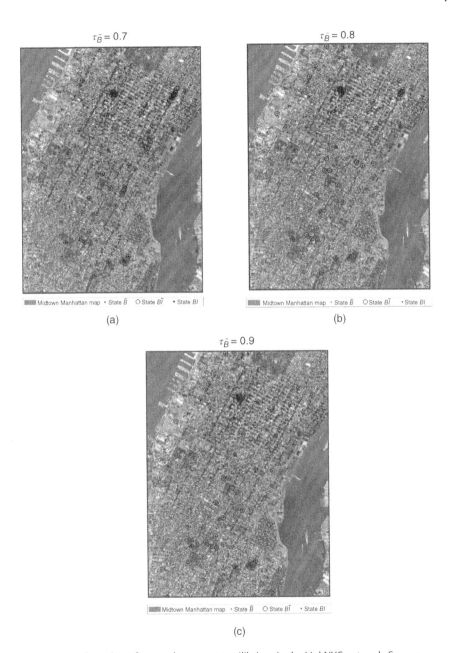

Figure 8.11 Snapshot of network states at equilibrium in the LinkNYC network. Source: LinkNYC network.

the equilibrium is achieved. Again, the experiment is repeated for different target thresholds for bot-free population, i.e. $\tau_{\bar{B}} = 0.7, 0.8$, and 0.9. The snapshots of the network states at equilibrium are shown in Figure 8.11. A similar behavior is observed as the network increasingly becomes bot free at equilibrium as the patching rates are increased. The time evolution of un-compromised devices for each of the thresholds is recorded in Figure 8.10b. We start off with infecting around 40% of the devices with malware and allow it to spread. It results in an infection of around 92% of the network with 8% un-compromised devices. However, once the patching policy is implemented, the network recovers sharply and is able to achieve much higher bot-free proportions than the target. It is pertinent to mention that since the network is not a PPP, the spread of malware is more difficult. Hence, the developed patching policy is more effective than expected, resulting in better performance of the policy. Therefore, a Poisson network assumption proves to be a more conservative approximation of the real network, which is favorable in practice as the results correspond to a worst case scenario.

8.4   Summary and Conclusion

In this chapter, we develop a mathematical model to study the formation of botnets in wireless IoT networks. A customized dynamic population process model coupled with a PPP based network model is proposed to capture the evolution of different types of population in the network while keeping the network geometry into account. The proposed model characterizes the behavior of malware transmission from one device to another using the wireless interface along with the propagation of control commands between bot devices in the network. A network defender is assumed to patch the devices to avert the formation of a botnet that may trigger a coordinated attack at a later stage. The equilibrium state of malware infection and message propagation in the devices is determined using approximate analysis. The results are then used to develop a network defense problem that aims to obtain optimal patching rates while minimizing the disruption to regular network operation under tolerable botnet activity. While the optimal patching problem may be non-convex, a dual decomposition algorithm with appropriate conditions is proposed to solve the optimization problem resulting in the optimal patching schedule for network devices based on their connectivity profile.

Part V

Resource Provisioning Mechanisms

9

Revenue Maximizing Cloud Resource Allocation

Cloud computing is becoming an essential component in the emerging Internet of things (IoT) paradigm. The available resources at the cloud such as computing nodes, storage, databases, etc. are often packaged in the form of virtual machines (VMs) to be used by remotely located IoT client applications for computational tasks. However, the cloud has a limited number of VMs available and hence, for massive IoT systems, the available resources must be efficiently utilized to increase productivity and subsequently maximize revenue of the cloud service provider (CSP). IoT client applications generate requests with computational tasks at random times with random complexity to be processed by the cloud. The CSP has to decide whether to allocate a VM to a task at hand or to wait for a higher complexity task in the future. We propose a threshold-based mechanism to optimally decide the allocation and pricing of VMs to sequentially arriving requests in order to maximize the revenue of the CSP over a finite time horizon. Moreover, we develop an adaptive and resilient framework that can counter the effect of real-time changes in the number of available VMs at the cloud server, the frequency, and the nature of arriving tasks on the revenue of the CSP.

9.1 Cloud Service Provider Resource Allocation Problem

We consider a CSP having a set of $N_t \in \mathbb{Z}^+$ available VMs at time $t \in [0, T]$. The VMs are identical and are characterized by their computational efficiency[1] denoted by $q \in [0, 1]$. The CSP receives requests[2] for computation by IoT client applications. These requests arrive sequentially at the cloud server according

1 The computational efficiency can be determined by evaluating the relative time taken by the VM to successfully execute a benchmark task.
2 Throughout the chapter, we use the word "requests" to refer to computational tasks generated by IoT applications that arrive at the CSP for processing.

Resource Management for On-Demand Mission-Critical Internet of Things Applications, First Edition.
Junaid Farooq and Quanyan Zhu.
© 2021 John Wiley & Sons, Inc. Published 2021 by John Wiley & Sons, Inc.

to a Poisson process with density $\lambda \in \mathbb{R}^+$ requests per unit time. Each task has computational complexity denoted by $X \in \mathbb{R}^+$. The computational complexity can be measured in terms of the number of CPU cycles or equivalently the time required to complete a given computational task. The computational complexities of sequentially arriving tasks are considered to be independent and identically distributed (i.i.d.) random variables with probability density function (pdf) and cumulative distribution function (cdf) denoted by $f_X(x)$ and $F_X(x)$, respectively.

The utility of the ith arriving client application with a task complexity of $x_i, i \in \mathbb{Z}^+$, which is allocated a VM with efficiency q, is measured by the product qx_i, which refers to the resulting value created by the allocation or the productivity. Since the available VMs are limited, the CSP needs to allocate the VMs to only the high complexity arriving tasks in order to increase the total productivity of the clients as well as efficient utilization of available computational resources. Creating higher value sets the ground for the CSP to charge higher prices and hence generate more revenue. However, the decision has to be taken immediately[3] upon arrival of the tasks without knowledge of tasks arriving in the future. Therefore, the CSP has the option to either allocate one of the available VMs or to refuse the requesting application.

9.2 Allocation and Pricing Rule

In order to allocate available VMs to randomly arriving computational requests, we adapt the result from the sequential stochastic assignment literature, which is based on the Hardy–Littlewood–Polya inequality [60] and is stated by the following theorem:

Theorem 9.1 (*Adapted from [3]*) *If there are n VMs with computational efficiencies $q_1, q_2, \ldots, q_n$ such that $0 < q_1 \leq q_2 \leq \cdots \leq q_n$, then there exists a set of functions*

$$0 = z_{n+1}(t) \leq z_n(t) \leq \cdots \leq z_1(t) \leq z_0(t) = \infty. \tag{9.1}$$

such that it is optimal (in terms of social welfare) to assign a VM with efficiency q_i to an incoming task with complexity x if $z_{n-i+1}(t) \leq x \leq z_{n-i}(t)$. Furthermore, if $x < z_n(t)$, it is optimal not to allocate it.

In the case of identical objects, the CSP needs to set only a single dynamic threshold, which we refer to as the *qualification threshold*, that allows it to decide whether to allocate a VM or not based on the nature of the arriving task.

---

3 We assume that the tasks are impatient and need to be processed immediately without delay.

Let $y_{N_t}(t) \in \mathbb{R}^+$, $\forall t \in [0, T]$ denote the threshold if N_t VMs are available for allocation at time t. In other words, only the requests with $x_i \geq y_{N_t}(t)$, $i \in \mathbb{Z}^+$ will be allocated to an available VM at time t. The allocation process has to be completed within a finite time horizon denoted by T. Since the VMs have no commercial value if they remain idle or unallocated during the allocation period, therefore the threshold needs to be dynamic in order to efficiently generate revenue from available resources. The decision problem lies in that fact that it may be more valuable to assign a VM to a low complexity task than waiting for a high complexity task to arrive in the future, which may not ever realize.

The next step is to develop a pricing scheme for the available VMs. Since all the VMs are identical in terms of their performance, therefore they must be priced equally. The threshold based allocation policy provides a natural method for pricing the available VMs. Since each arriving task that is successfully allocated a VM at time t receives a value of at least $qy_{N_t}(t)$, therefore, it is fair to charge the price $\mathcal{P} : [0, 1] \times \mathbb{R}^+ \times [0, T] \to \mathbb{R}^+$ to a qualified task for being processed by a VM as follows:

$$\mathcal{P}(q, y_{N_t}(t), t) = qy_{N_t}(t) + S(t), \tag{9.2}$$

where $S(t)$ represents the constant additional pricing independent of the allocation. This pricing policy is implementable as any individually rational client will be willing to pay at least an amount equal to its received value. Note that $S(t)$ can be used to adjust the prices due to external factors such as promotions, packages, pricing agreements, etc. In Section 9.3, we provide the dynamically optimal allocation threshold and the resulting price charged by the CSP to allocated tasks.

9.3 Dynamic Revenue Maximization

We will begin by quantifying the total expected revenue of the CSP and subsequently derive the optimal dynamic threshold that maximizes the revenue. The total expected revenue of the CSP $\mathcal{R} : [0, 1] \times \mathbb{R}^+ \times [0, T] \to \mathbb{R}^+$ if N_t identical VMs with computational efficiency q are available at time t and a qualification threshold $y_{N_t}(t)$ is used from time t onwards can be expressed as follows:

$$\mathcal{R}(q, y_{N_t}(t), t) = \sum_{n=1}^{N_t} q \int_t^T y_{N_t}(s) h_n(s) ds + \kappa_T, \tag{9.3}$$

where $h_n(t)$ is the density of waiting time until the nth arrival of a qualifying task, i.e. having a task complexity greater than $y_{N_t}(t)$, and κ_T is a constant factor due to the additional pricing function $S(t)$. The density of waiting time can be expressed by the density of the nth arrival in a non-homogeneous Poisson process with intensity $\hat{\lambda}(s) = \lambda(1 - F(y_{N_t}(s)))$. Consequently, the density can be written as follows [114]:

$$h_n(s) = \hat{\lambda}(s) \exp\left(-\int_t^s \hat{\lambda}(u)du\right) \frac{\left(\int_t^s \hat{\lambda}(u)du\right)^{n-1}}{(n-1)!}, \quad t \leq s \leq T. \tag{9.4}$$

The objective is to select a time-varying threshold $y_{N_t}(t)$ that maximizes the expected revenue functional given by (9.3). The problem can be formally stated as follows:

$$\underset{y_{N_t}(t)}{\text{maximize}} \quad \mathcal{R}(q, y_{N_t}(t), t)$$

$$\text{subject to} \quad y_{N_t}(t) \geq 0, \quad \forall t \in [0, T].$$

Note that the optimization is over the space of functions where an optimal function $y_{N_t}(t)$ is sought for a given number of available VMs at time t. Our aim is to design the threshold function that strikes the optimal balance between the number of qualifying tasks and the generated revenue. In the sequel, we provide the optimal qualification threshold for maximum revenue generation by the CSP and the properties of the optimal policy.

Theorem 9.2 *If N_t VMs are available to the CSP at time t, computational requests arrive sequentially according to a Poisson process with intensity λ and the computational complexity of tasks are i.i.d. random variables with pdf $f_X(x)$ and cdf $F_X(x)$, then it is optimal to allocate an available VM to an incoming computational request if the complexity of an upcoming task $x \geq y_{N_t}(t)$. The optimal $y_{N_t}(t)$ satisfies the following integral equation:*

$$y_{N_t}(t) = \frac{1 - F_X(y_{N_t}(t))}{f_X(y_{N_t}(t))} + \lambda \int_t^T \frac{(1 - F_X(y_{N_t}(s)))^2}{f_X(y_{N_t}(s))} J_{N_t}(t, s)ds, \tag{9.5}$$

where $J_{N_t}(t, s)$ can be expressed as follows:

$$J_{N_t}(t, s) = \frac{1}{\sum_{n=1}^{N_t} \frac{1}{(n-1)!} \left(\int_t^s \hat{\lambda}(u)du\right)^{n-1}}$$

$$\times \sum_{n=1}^{N_t} \frac{1}{(n-1)!} \left(\left(\int_t^s \hat{\lambda}(u)du\right)^{n-1} - (n-1)\left(\int_t^s \hat{\lambda}(u)du\right)^{n-2}\right).$$

$$\tag{9.6}$$

Proof: Let $H(s) = \int_t^s \hat{\lambda}(u)du = \int_t^s \lambda(1 - F_X(y_{N_t}(u)))du$. Then the expected revenue at time t if N_t VMs are available can be written as follows:

$$\mathcal{R}(\{q\}_{N_t}, t) = q \int_t^T \sum_{n=1}^{N_t} \frac{1}{(n-1)!} F_X^{-1}\left(1 - \frac{H'(s)}{\lambda}\right) H'(s)e^{-H(s)}(H(s))^{n-1}ds + \kappa_T.$$

$$\tag{9.7}$$

This functional can be optimized for the time varying threshold $y_{N_t}(t)$ using the calculus of variations [50]. We denote the kernel of integration as

$$\mathcal{L}(s, H(s), H'(s)) = \sum_{n=1}^{N_t} \frac{H'(s)e^{-H(s)}H^{n-1}(s)}{(n-1)!} F_X^{-1}\left(1 - \frac{H'(s)}{\lambda}\right).$$ (9.8)

The Euler–Lagrange equation [50] represents the necessary condition satisfied by $H(s)$ to be a stationary function of the expected revenue $R(\{q\}_{N_t}, t)$ and can be written as follows:

$$\frac{\partial \mathcal{L}(s, H(s), H'(s))}{\partial H(s)} - \frac{d}{dt}\frac{\partial \mathcal{L}(s, H(s), H'(s))}{\partial H'(s)} = 0.$$ (9.9)

The partial derivatives and the condition satisfied by the resulting Euler–Lagrange equation are given by (9.10), (9.11), and (9.12) as follows:

$$\frac{\partial \mathcal{L}(s, H(s), H'(s))}{\partial H(s)} = \mathcal{L}_H = \sum_{n=1}^{N} \frac{1}{(n-1)!} e^{-H(s)} F_X^{-1}$$

$$\times \left(1 - \frac{H'(s)}{\lambda}\right) H'(s)((n-1)H^{n-2}(s) - H^{n-1}(s)),$$ (9.10)

$$\frac{d}{dt}\mathcal{L}_{H'} = \sum_{n=1}^{N} \frac{1}{(n-1)!} \left[-2\frac{H^{n-1}(s)H''(s)}{\lambda f_X\left(F_X^{-1}\left(1 - \frac{H'(s)}{\lambda}\right)\right)} - H'(s)H^{n-1}(s)F_X^{-1} \right.$$

$$\times \left(1 - \frac{H'(s)}{\lambda}\right) + (n-1)H^{n-1}(s)H'(s) \times F_X^{-1}\left(1 - \frac{H'(s)}{\lambda}\right)$$

$$\left. - \frac{H^{n-1}(s)(H'(s))^2}{\lambda F_X^{-1}\left(1 - \frac{H'(s)}{\lambda}\right)} - \frac{H^{n-1}(s)H'(s)H''(s)f_X'(F_X^{-1}(1 - \frac{H'(s)}{\lambda}))}{\lambda^2 f_X^3\left(F_X^{-1}\left(1 - \frac{H'(s)}{\lambda}\right)\right)} \right].$$ (9.11)

$$\sum_{n=1}^{N} \frac{1}{(n-1)!} \left[2H^{n-1}(s)H''(s) - (H'(s))^2(H^{n-1}(s) - (n-1)H^{n-2}(s)) \right.$$

$$\left. + \frac{H^{n-1}(s)H'(s)H''(s)f_X'(F_X^{-1}(1 - \frac{H'(s)}{\lambda}))}{\lambda f_X^2\left(F_X^{-1}\left(1 - \frac{H'(s)}{\lambda}\right)\right)} \right] = 0.$$ (9.12)

The expression in (9.12) can be further reduced as follows:

$$
2H''(s) - (H'(s))^2 \frac{\sum_{n=1}^{N_t} \frac{1}{(n-1)!}(H^{n-1}(s) - (n-1)H^{n-2}(s))}{\sum_{n=1}^{N_t} \frac{1}{(n-1)!}H^{n-1}(s)}
$$

$$
+ \frac{H'(s)H''(s)f'_X(F_X^{-1}(1 - \frac{H'(s)}{\lambda}))}{\lambda f_X^2 \left(F_X^{-1}\left(1 - \frac{H'(s)}{\lambda}\right)\right)} = 0. \tag{9.13}
$$

Let $J_{N_t}(t,s) = \frac{\sum_{n=1}^{N_t} \frac{1}{(n-1)!}(H^{n-1}(s) - (n-1)H^{n-2}(s))}{\sum_{n=1}^{N_t} \frac{1}{(n-1)!}H^{n-1}(s)}$. Then, plugging back $H(s) = \int_t^s \lambda(1 - F_X(y_{N_t}(u)))du$ results in the following:

$$
-2y'_{N_t}(s) - \frac{\lambda(1 - F_X(y_{N_t}(s)))^2 J_{N_t}(t,s)}{f(y_{N_t}(s))} - \frac{(1 - F_X(y_{N_t}(s)))y'_{N_t}(s)f'_X(y_{N_t}(s))}{f_X^2(y_{N_t}(s))} = 0 \tag{9.14}
$$

It can be further expressed as follows:

$$
-y'_{N_t}(s) - y'_{N_t}(s)\left(1 + \frac{(1 - F_X(y_{N_t}(s)))f'_X(s)}{(f_X(y_{N_t}(s)))^2}\right) = \frac{\lambda(1 - F_X(y_{N_t}(s)))^2 J_{N_t}(t,s)}{f_X(y_{N_t}(s))}, \tag{9.15}
$$

Since $\frac{d}{ds}\left(\frac{1 - F_X(y_{N_t}(s))}{f(y_{N_t}(s))}\right) = -y'_1(s)\left(1 + \frac{(1 - F_X(y_{N_t}(s)))f'_X(y_{N_t}(s))}{(f_X(y_{N_t}(s)))^2}\right)$, so the condition in (9.15) can be written as follows:

$$
\frac{d}{ds}\left(\frac{1 - F_X(y_{N_t}(s))}{f_X(y_{N_t}(s))}\right) = y'_1(s) + \lambda \frac{(1 - F_X(y_{N_t}(s)))^2}{f_X(y_{N_t}(s))}J_{N_t}(t,s). \tag{9.16}
$$

Integrating both sides with respect to s from t to T results in the following:

$$
\left(\frac{1 - F_X(y_{N_t}(T))}{f_X(y_{N_t}(T))}\right) - \left(\frac{1 - F_X(y_{N_t}(t))}{f_X(y_{N_t}(t))}\right) = y_{N_t}(T)
$$

$$
- y_{N_t}(t) + \lambda \int_t^T \frac{(1 - F_X(y_{N_t}(s)))^2}{f_X(y_{N_t}(s))}J_{N_t}(t,s)ds. \tag{9.17}
$$

Using the boundary condition, $y_{N_t}(T) = \frac{1 - F_X(y_{N_t}(T))}{f(y_{N_t})(T)}$, i.e. at the terminal time only the virtual valuation of the users can be recovered, it follows that the cutoff curve $y_{N_t}(t)$ satisfies the equation given by Theorem 9.2. $\qquad \square$

The behavior of the optimal dynamic threshold for large number of available VMs is provided by the following corollary.

Corollary 9.1 *If the number of available VMs is large, then the revenue maximizing threshold becomes constant and the allocation mechanism reduces to a first price auction mechanism, i.e. allocate a VM to a task if the complexity is higher than the virtual valuation.*

Proof: In the optimal allocation policy, if we let $N_t \to \infty$, then the optimal threshold solves the following integral equation:

$$y_\infty(t) = \frac{1 - F_X(y_\infty(t))}{f_X(y_\infty(t))} + \int_t^T \frac{(1 - F_X(y_\infty(t)))^2}{f_X(y_\infty(t))} \left(\lim_{N_t \to \infty} J_{N_t}(t, s) \right) ds.$$

$$(9.18)$$

Now, $\lim_{N_t \to \infty} J_{N_t}(t, s)$ can be evaluated as follows:

$$\lim_{N_t \to \infty} J_{N_t}(t, s) = \frac{\sum_{n=1}^{\infty} \frac{H^{n-1}(s)}{(n-1)!} - \sum_{n=1}^{\infty} \frac{(n-1)H^{n-2}(s)}{(n-1)!}}{\sum_{n=1}^{\infty} \frac{H^{n-1}(s)}{(n-1)!}}$$

$$= \frac{e^{H(s)} - e^{H(s)}}{e^{H(s)}} = 0.$$

$$(9.19)$$

Therefore, it follows that $y_\infty(t) = \frac{1 - F_X(y_\infty(t))}{f_X(y_\infty(t))}$. Note that $x - \frac{1 - F_X(x)}{f_X(x)}$ is referred to as the virtual valuation of the agent of type x in mechanism design literature [101]. Hence, it can be concluded that if the number of available VMs is large, then only the virtual valuation of the arriving tasks can be recovered and the CSP is willing to offer the VMs for lowest possible threshold. □

The behavior of the dynamically optimal qualification threshold with a variation in the number of available VMs at time t can be summarized by the following theorem.

Theorem 9.3 *The qualification threshold of the tasks and consequently price of VMs decreases as the number of available VMs at the cloud server increases and vice versa, i.e. $y_{M_t}(t) \leq y_{N_t}(t)$ if $M_t \geq N_t, \forall t$.*

Proof: First we need to show that for $\{M_t, N_t \in \mathbb{Z}^+ : M_t \geq N_t\}$, $J_{M_t}(t, s) \leq J_{M_t}(t, s), \forall t, s$. To do this we will show that $J_{N_t+1}(t, s) \leq J_{N_t}(t, s)$. It is equivalent to showing that $J_{N_t+1}(t, s) - J_{N_t}(t, s) \leq 0$, i.e.

$$\frac{\sum_{n=1}^{N_t+1} \frac{H^{n-1}(s)}{(n-1)!} - \sum_{n=1}^{N_t+1} \frac{(n-1)H^{n-2}(s)}{(n-1)!}}{\sum_{n=1}^{N_t+1} \frac{H^{n-1}(s)}{(n-1)!}} - \frac{\sum_{n=1}^{N_t} \frac{H^{n-1}(s)}{(n-1)!} - \sum_{n=1}^{N_t} \frac{(n-1)H^{n-2}(s)}{(n-1)!}}{\sum_{n=1}^{N_t} \frac{H^{n-1}(s)}{(n-1)!}} \leq 0.$$

$$(9.20)$$

It can be further expressed as follows:

$$\left(\sum_{n=1}^{N_t} \frac{H^{n-1}(s)}{(n-1)!} \right) \left(\frac{H^{N_t}(s) - N_t H^{N_t-1}(s)}{N_t!} \right)$$

$$- \left(\sum_{n=1}^{N_t} \frac{H^{n-1}(s) - (n-1)H^{n-2}(s)}{(n-1)!} \right) \frac{H^{N_t}(s)}{N_t!} \leq 0. \tag{9.21}$$

Expanding the condition results in the following:

$$\sum_{n=1}^{N_t} \left(\frac{H^{N_t+n-1}(s) - N_t H^{N_t+n-2}(s) - H^{N_t+n-1}(s)}{(n-1)!} + \frac{(n-1)H^{N_t+n-2}(s)}{(n-1)!} \right) \leq 0. \tag{9.22}$$

It is equivalent to the following condition:

$$\sum_{n=1}^{N_t} \frac{H^{N_t+n-2}(s)(n+1-N_t)}{(n-1)} \leq 0, \tag{9.23}$$

which is true since $(n + 1 - N_t) \leq 0, \forall n = 1, \ldots, N_t$. Therefore, it is evident that $J_{N_t+1}(t, s) \leq J_{N_t}(t, s)$. Using induction it can be shown that the inequality $J_{M_t}(t, s) \leq J_{N_t}(t, s)$ holds for general M_t and N_t such that $M_t \geq N_t, \forall t$. From Theorem 9.2, the result follows directly with the assumption of increasing virtual valuations, i.e. $x - \frac{1-F_X(x)}{f_X(x)}$ is increasing in x. $\square$

In Section 9.3.1, we discuss how the dynamically optimal policy leads to an adaptive and resilient behavior in the revenue of the CSP and describe the developed mechanism algorithmically.

9.3.1 Adaptive and Resilient Allocation and Pricing Policy

The number of available VMs at the cloud server may change over time as some of them might become unavailable due to failure or malicious attacks [64]. The CSP might also destroy the created VMs in order to free up computational resources for other applications. On the other hand, previously allocated VMs might be released by applications or new VMs may be provisioned by the CSP in real-time to accommodate higher demand. However, a change in the available number of VMs may affect the expected revenue of the CSP under a particular allocation and pricing policy particularly if there is a significant decrease in the number of remaining VMs. In order to reduce any negative impact on the expected revenue of the CSP, the proposed revenue maximizing framework will react to the changes in the available number of VMs by adapting the qualification threshold or equivalently the price.

Furthermore, the developed framework can react to changes in the frequency and the nature of computational requests. The optimal resilient policy

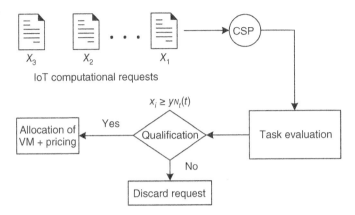

Figure 9.1 Flow diagram of the adaptive and resilient resource allocation and pricing mechanism.

is denoted by $\Pi(\tilde{\lambda}_t, \tilde{N}_t, t)$, where $\tilde{\lambda}_t$ is the rate of arrival of requests and $\tilde{N}_t$ represents the number of available VMs at time t. Note that $\tilde{N}_t = N_t + \eta_t$, where $\eta_t \in \mathbb{Z}$ is the change in the number of available VMs at time t. The adaptive qualification threshold $\tilde{y}_{N_t}(t)$ can be pre-computed using the optimal policy framework presented earlier in this section. The policy then becomes a lookup table that the CSP uses to allocate and price the upcoming tasks. Note that the variations in the inputs of the framework can be directly incorporated into the derived results. For instance, if η_t VMs enter/leave the system at time t, then it is equivalent to as if there were $N_t + \eta_t$ available VMs at time t. Hence, the optimal threshold corresponding to $N_t + \eta_t$ must be used for time t onwards for maximizing revenue.

The algorithm proceeds as follows. While the allocation period has not expired and there is still an available VM at the CSP, if an IoT application requests for computation, then the first step is to evaluate the task complexity. Once the complexity is determined, it is compared against the decision threshold. However, the optimal threshold used will depend on the current situation at the CSP. Hence, the updated number of available VMs $\tilde{N}_t$ and the updated arrival rate of requests $\tilde{\lambda}$ is used to read off the optimal policy from the lookup table $\Pi(\tilde{\lambda}_t, \tilde{N}_t, t)$ at time t. A flow diagram is provided in Figure 9.1 to illustrate the sequence of the mechanism.

9.4 Numerical Results and Discussions

In this section, we provide numerical results for the proposed adaptive and resilient optimal dynamic allocation and pricing framework. We assume a single CSP having N_t available VMs to allocate to arriving computational requests within an allocation time horizon of $T = 12$ hours. The number of available VMs available at time $t = 0$, referred to as N_0, is set to be 100. The computational efficiency of the VMs is selected to be $q = 1$ without loss of generality. Note that the characteristics of the VMs are only relevant to the pricing policy but not the allocation.

The tasks arrive at the CSP according to a homogeneous Poisson process with intensity $\lambda = 100$ requests per hour unless otherwise stated. We also assume that the complexity of sequentially arriving computational requests is distributed according to an exponential distribution with a mean of $\frac{1}{\alpha}$, i.e. $f_X(x) = \alpha e^{-\alpha x}$, and $F_X(x) = 1 - e^{-\alpha x}$. For simplicity, we select $\alpha = 1$, resulting in an average task complexity of 1. The optimal task qualification thresholds in this case if N_t VMs are available at time t can be obtained by the solution of the integral equation expressed by (9.24):

$$
y_{N_t}(t) = \frac{1}{\alpha} + \frac{\lambda}{\alpha} \int_t^T e^{-\alpha y_{N_t}(s)}
$$

$$
\times \frac{\sum_{n=1}^{N_t} \frac{1}{(n-1)!} \left[\left(\int_t^s \lambda e^{-\alpha y_{N_t}(s)} du \right)^{n-1} - (n-1)\left(\int_t^s \lambda e^{-\alpha y_{N_t}(s)} du \right)^{n-2} \right]}{\sum_{n=1}^{N_t} \frac{1}{(n-1)!} \left(\int_t^s \lambda e^{-\alpha y_{N_t}(s)} du \right)^{n-1}} ds.
$$

$$(9.24)$$

The equation can be solved numerically using the Picard fixed point iteration [18]. Figure 9.2 shows the dynamic thresholds for qualification of an arriving task for low ($\lambda = 10$) and high ($\lambda = 100$) arrival rates of the requests. It can be observed in general that the qualification thresholds decrease as the time approaches toward the terminal time. This is due to the fact that the valuable option of allocating an available VM to a higher complexity task reduces in probability. Furthermore, as we approach the horizon, it is more valuable to allocate a VM to a lower complexity task than to not allocate it at all. It can also be observed from Figure 9.2a that for lower arrival rates, the thresholds drop quickly as compared with the thresholds for the high arrival rates in Figure 9.2b. This is because the expected arrivals are lower in the former and hence the mechanism adjusts the thresholds to qualify more arrivals to tap the revenue potential. The associated pricing curves follow a similar trend as the allocation thresholds except that they are scaled by the characteristics of the VMs. However, in the considered situation, they are identical since $q = 1$.

Next we investigate the adaptive and resilient behavior of the proposed mechanism. A set of failures and capacity enhancements are simulated at fixed times. For instance, a loss of 15 VMs and 5 VMs is assumed to occur at $t = 2$ hours and $t = 6$ hours, respectively. Similarly, new additions of 10 VMs and 5 VMs are assumed to occur at $t = 4$ hours and $t = 8$ hours. Note that when a loss of 15 VMs occurred at $t = 2$ hours, the situation becomes

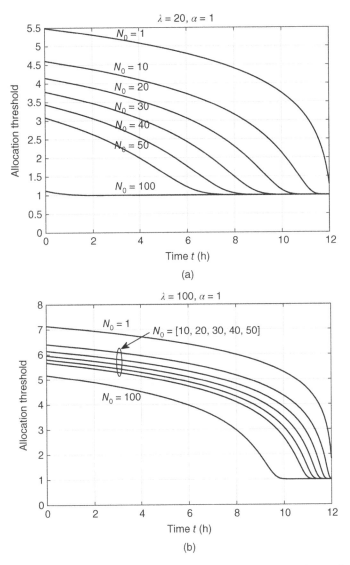

Figure 9.2 Optimal allocation thresholds for low and high arrival rates of computational requests. (a) Low frequency of requests. (b) High frequency of requests.

equivalent to as if initially the CSP had 85 available VMs. Therefore from $t = 2$ onwards, the optimal revenue maximizing policy is to use the threshold and pricing corresponding to the $N_0 = 85$ curve. As the number of available VMs change over time, the optimal policy needs to be updated. The optimal dynamic allocation policy for the aforementioned events is shown by the bold

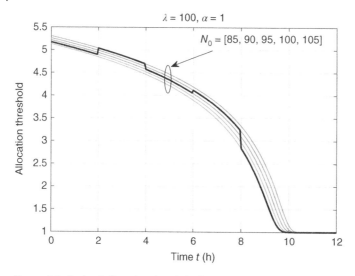

Figure 9.3 Optimal allocation threshold for varying number of available VMs.

line in Figure 9.3. Notice that as the number of available VMs decreases, the allocation threshold and the price jumps in order to make up for the lost revenue. Similarly, if new VMs become available, the threshold and prices drop in order to strike a new balance between the qualifying tasks and the payment.

In Figure 9.4, we show the behavior of the adaptive and resilient mechanism on the expected revenue of the CSP in response to the variations in the number

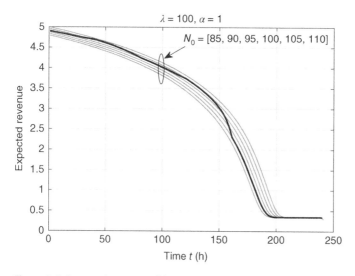

Figure 9.4 Expected revenue of the CSP over time under variations in the number of available VMs.

of available VMs at the cloud server. It can be observed that the adaptive strategy is able to maintain a high expected revenue despite variations in the number of available VMs. Note that during the times when there is a drop in the number of available VMs, the expected revenue does not fall as much due to a rectified allocation and pricing policy as illustrated in Figure 9.3. Hence, it is shown that a timely rectification of allocation and pricing decision enables the mechanism to be adaptive and resilient against any significant changes in the available resources to the CSP.

9.5 Summary and Conclusion

In this chapter, we have proposed an adaptive and resilient dynamic revenue maximizing framework for cloud computing environments. The framework uses a threshold-based filtering policy making real-time allocation and pricing decisions for sequentially arriving computational requests. It has been shown that the framework is adaptive and resilient to changes in the number of available VMs or the statistical properties of the arrivals. The set of optimal policies can be pre-computed and used as a lookup table as conditions at the cloud server change over time. Therefore, the developed framework provides an optimal and implementable mechanism for allocation and pricing in cloud computing environments. Future directions in this line of work may include developing optimal allocation policies for multiple types of identical resources available at the CSP. Furthermore, the allocation framework can be extended to multiple layers such as in fog/edge computing paradigms.

10

Dynamic Pricing of Fog-Enabled MC-IoT Applications

Most existing work on job dispatching and scheduling is focused on the static assignment of tasks to computational resources in the fog-cloud environment. For instance, a task offloading framework using matching theory has been proposed in [25]. However, it is based on static pairing between tasks and fog nodes and the dynamic aspects of task arrival are not considered. Similarly, an index based task assignment and scheduling of tasks is proposed in [56]. Several works do consider real-time task allocation and dispatch. For instance, an online job dispatching and scheduling algorithm is proposed in [134], where jobs are released in arbitrary order and times by mobile devices and offloaded to unrelated servers with both upload and download delays. However, the approach is based on a greedy algorithm and is not strategic, i.e. does not make use of the statistical information about computational requests to make more effective allocation decisions. Furthermore, pricing and revenue maximization have been looked into sparingly.

In this chapter, we present a revenue maximizing perspective toward allocation and pricing in fog based systems designed for mission-critical IoT (MC-IoT) applications. The quality-of-experience (QoE) resulting from the pairing of fog resources with computation requests is used as a basis for pricing. We develop a dynamic policy framework leveraging the literature in economics, mechanism design [51], and dynamic revenue maximization [52] to provide an implementable mechanism for dynamic allocation and pricing of sequentially arriving Internet of things (IoT) requests that maximize the expected revenue of the cloud service provider (CSP). The developed optimal policy framework assists in both determining which fog node to allocate an incoming task to and determining the price that should be charged for it for revenue maximization. The proposed policy is statistically optimal, dynamic, i.e. adapts with time, and is implementable in real-time as opposed to other static matching schemes in the literature. The dynamically optimal solution can be computed offline and implemented in real-time for sequentially arriving computation requests.

Resource Management for On-Demand Mission-Critical Internet of Things Applications, First Edition.
Junaid Farooq and Quanyan Zhu.

10.1 Edge Computing and Delay Modeling

We consider a CSP in a fog enabled IoT ecosystem having a set of k available fog nodes in addition to the main cloud server and serving a certain geographical region containing MC-IoT devices. The fog nodes have an associated average latency denoted by $l_i \in \mathbb{R}, i \in \{1, \ldots, k\}$, which depends on the distances of the fog nodes from the locations of origin of the processing requests. The number of available virtual machine instances (VMIs) at the fog nodes is denoted by $n_i, i \in \{1, \ldots, k\}$ and each of the available VMI is characterized by its processing delay for a fixed number of computational operations denoted by $\tau_{ij}^{(p)}, i = \{1, \ldots, k\}, j \in \{1, \ldots, n_i\}$. There are a total of $\sum_{i=1}^{k} n_i = N$ VMIs available for allocation by the CSP to sequentially arriving MC-IoT requests. The latency of each individual VMI is denoted by $\tau_{ij}^{(l)} = l_i, \forall j = \{1, \ldots, n_i\}$. The end-to-end delay offered by the VMIs can be evaluated as $\tau_{ij}^{(l)} + \tau_{ij}^{(p)} + \tau^{(o)}$, where $\tau^{(o)}$ represents the other delays including the transmission delay over the air interfaces. Consequently, the average response rate of the jth VMI at compute node i can be expressed as follows:[1]

$$r_{ij} = \frac{1}{\tau_{ij}^{(l)} + \tau_{ij}^{(p)} + \tau^{(o)}}. \tag{10.1}$$

With an abuse of notation, the average response rates of the available VMIs can further be denoted by $\mathbf{r} = [r_1, r_2, \ldots, r_N]$, where $r_1 \geq r_2 \geq \cdots \geq r_N$, and $\mathcal{M}$ defines the mapping of values of the pair $(i, j), i \in \{1, \ldots, k\}, j \in \{1, \ldots, n_i\}$ from the original r_{ij} to the set $n = \{1, \ldots, N\}$ in the new response rates denoted by r_n. It implies that the VMI corresponding to a response rate r_1 is the best in terms of end-to-end delay for processing IoT data, while the one corresponding to r_N is the least favorable in terms of end-to-end delay.

MC-IoT devices in a given geographical region are connected to the CSP via a set of IoT gateways. We assume that requests for processing by MC-IoT applications arrive sequentially at the CSP according to a homogeneous Poisson process with intensity λ. Each request is characterized by its maximum tolerable delay for successful operation denoted by $d_i, i \geq 1$. In other words, there is a minimum required response rate by the applications. Upon arrival of a request by an MC-IoT application, the CSP has to make a decision to forward the data to one of available VMIs at the fog nodes. An illustration of the system model is provided in Figure 4.3.

It is pertinent to mention that before the actual task offloading, the task metadata is sent to the cloud server for the allocation decision. Furthermore, we assume that the allocation decision is instantaneous and the delay in transmitting metadata is not relevant in the computation of delay experienced in the

---

1 Performance metrics and utility of similar forms are used in the literature for fog-enabled IoT systems, e.g. [25, 154].

transmission and processing of actual application data since it occurs only once before the allocation is made.

10.2   Allocation Efficiency and Quality of Experience

Upon arrival of a request by an MC-IoT application, the CSP determines the sensitivity of the application to a delay in the response and is subsequently required to allocate it to one of the available VMIs at the fog nodes while charging a particular price. The VMIs are allocated to requesting applications for long term, and therefore, once a VMI is allocated to an application, it becomes unavailable for allocation to other applications in the future.[2] Since the VMIs are perishable, i.e. they have no value if not successfully allocated within the allocation time horizon T, so it is important for the CSP to optimally allocate the available resources within a certain time frame.[3] It also motivates the idea of dynamic pricing, i.e. to charge higher prices earlier and reduce them as time goes on for maximizing the revenue of the CSP.

Let $x_i = d_i^{-1}$ be the minimum required response rate by the ith application. Each application reports this characteristic to the CSP at the time of requesting service. We assume that each $x_i \in \mathbb{R}, i \geq 1$ is an independent and identically distributed (i.i.d.) random variable with a probability density function (pdf) denoted by $f_X(x)$ and a cumulative distribution function (cdf) denoted by $F_X(x)$.[4] The product $x_i r_j$ can be used as a measure of efficiency when the ith application is allocated to the jth available VMI. The QoE as a result of the pairing, denoted by $\Phi(x_i, r_j)$, can be expressed as follows:

$$\Phi(x_i, r_j) = (x_i r_j)^{\frac{1}{\eta}}, \quad i, j \in \{1, \dots, N\}, \tag{10.2}$$

where $\eta \geq 1$ is a constant controlling the rate of increase in QoE with respect to the efficiency of allocation. The concave nature of the QoE function implies that an increase in allocation efficiency results in diminishing improvements in the QoE of the applications.

In this section, we provide a framework for dynamically maximizing the expected revenue of the CSP based on the QoE of the users. We focus on a mechanism design approach, which is based on developing optimal implementable policies for allocation and pricing. A direct mechanism is provided

---

2 Note that this assumption is only made for obtaining an optimal dynamic policy using an open loop methodology. In practical implementation, the policy can be adapted based on how many VMIs are available.

3 The time horizon refers to the period over which the allocation has to occur, which can be related to the demand window.

4 We assume that the probability distribution of the arriving tasks is known *a priori*. This is done for analytical tractability and policy development as data-driven approaches are prohibitive in terms of obtaining an implementable dynamic policy.

whereby each requesting application reports its minimum required response rate and the CSP allocates one of the available VMIs to it. For the optimal allocation to be implementable, an allocation policy and a payment rule are required that is incentive compatible[5] [101] in the presence of individually rational users. We first state an allocation rule that satisfies the aforementioned conditions and then provide a pricing strategy that subsequently implements the allocation.

10.2.1 Allocation Policy

Based on the required response rate of each randomly arriving computational request by the IoT devices, the CSP has to allocate a VMI at one of the fog nodes to maximize the expected QoE of the users. It is clear that an application that requires a high response rate should be allocated to the VMI offering a high response rate and vice versa for efficient utilization of resources. However, the problem rests in the fact that an allocation decision has to be made without knowledge of the type of applications that will request in the future. It is shown in the literature [3] that a dynamically efficient allocation policy in such situations can be achieved using a partition on the characteristic of the sequentially arriving agent. We provide a deterministic and Markovian allocation policy $\pi_t(x, \mathbf{r}_t)^6 : \mathbb{R} \rightarrow \mathbf{y}_t$, i.e. at each time t, a fixed non-random policy is used that only depends on the current time instant and set of available VMIs at time t. The key result in the allocation policy is provided by the following theorem.

Lemma 10.1 (***Adapted from*** [3]) *A deterministic and Markovian policy at time t, i.e. $\pi_t(x, \mathbf{r}_t)$, is implementable if there exists a set of functions $y_i(t)$, $i = 1, \ldots, N_t$, such that $0 < y_{N_t}(t) < y_{N_t-1}(t) < \cdots < y_1(t) < y_0(t) = \infty$. The allocation policy is such that $\pi_t(x, \mathbf{r}_t) = r_i$ if $x \in [y_i(t), y_{i-1}(t)]$ and $\pi_t(x, \mathbf{r}_t) = \emptyset$ if $x < y_{N_t}(t)$.*

The allocation policy depends on the nature of the requesting applications as well as the available VMIs at time t. Lemma 10.1 implies that if at time t, the requesting application has a required response rate of greater than $y_{N_t}(t)$, then the CSP will not allocate any compute node to the application as it aspires to save the VMIs for higher valued application requests in the future. However, if the required response rate is between $y_i(t)$ and $y_{i-1}(t)$, then the ith best available VMI is allocated to the requesting application. It can be observed that the number of partitions or cutoff curves depends on the set of available compute nodes

---

5 Incentive compatibility is a concept from mechanism design theory that ensures that no agent has an incentive to misreport its privately known characteristic.
6 Throughout the chapter, the subscript t is used to refer to the time dependence.

at time t. In the subsequent subsections, we provide a suitable pricing scheme associated with the aforementioned policy and provide the optimal time varying cutoff values for efficient allocation.

10.2.2 Pricing Policy

For the proposed allocation policy, there is a need to appropriately price the applications for their achieved QoE. We assume that the MC-IoT applications are individually rational, i.e. no application will be willing pay more than the QoE it achieves by using the allocated VMI. The partition based allocation policy described in Section 10.2.1 provides a natural mechanism for pricing the applications for their proposed allocations. Since an application with a required response rate $x \in [y_i(t), y_{i-1}(t)]$ at time t is allocated to a VMI with a response rate r_i, it has achieved an improvement in QoE of at least $(r_i^{\frac{1}{n}} - r_{i+1}^{\frac{1}{n}}) y_i^{\frac{1}{n}}(t)$ as compared with the next best allocation. Therefore, it must pay an equivalent price to be allocated to a VMI with response rate r_i as compared with the one with r_{i+1}. This process can be continued recursively to obtain implementable prices for each of the available VMIs. Therefore, the optimal prices can be completely determined by the implementation conditions. The price charged to an MC-IoT application that is allocated to a jth best VMI, i.e. $x \in [y_j(t), y_{j-1}(t))$, at time t can be expressed as follows:

$$P_j(\mathbf{r}_t, t) = \sum_{i=j}^{N_t} \left(r_i^{\frac{1}{n}} - r_{i+1}^{\frac{1}{n}} \right) y_i^{\frac{1}{n}}(t). \tag{10.3}$$

Note that $r_i > r_{i+1}, \forall i = 1, \ldots, N - 1$ due to the initial ordering. The pricing policy is progressive in a relative sense with the lowest QoE achieved as a reference. We provide a simple example to further elaborate the pricing policy. If there is a single VMI available, the price would be $P_1(\{r_1\}, t) = r_1^{\frac{1}{n}} y_1^{\frac{1}{n}}(t)$, which is equivalent to the QoE of the application to which the VMI was allocated. However, if there are two VMIs available, the price for the lower response rate VMI is set to be $P_2(\{r_1, r_2\}, t) = r_2^{\frac{1}{n}} y_2^{\frac{1}{n}}(t)$ and the price for the higher response rate VMI is set to be $P_1(\{r_1, r_2\}, t) = r_2^{\frac{1}{n}} y_2^{\frac{1}{n}}(t) + (r_1^{\frac{1}{n}} - r_2^{\frac{1}{n}}) y_1^{\frac{1}{n}}(t)$. Note that the price for the latter is simply the price of the former plus the improvement in QoE experienced by getting a VMI with response rate r_1 instead of r_2. In other words, the lowest response rate VMI is priced at the base price equivalent to the QoE achieved by its allocation. The next higher one is priced at the base price plus the improvement in QoE achieved as a result of being allocated a better available VMI.

10.3   Optimal Allocation and Pricing Rules

In this section, the goal is to maximize the expected revenue generated by the CSP using the pricing strategy developed in Section 10.1. We first begin with solving the problem for the case of a single available VMI and then generalize it to multiple VMIs using a recursive approach.

10.3.1   Single VMI Case

If only a single VMI is available to the CSP with a response rate r_1, then the expected revenue generated by its allocation within a time period T can be expressed as follows:

$$R(\{r_1\}, t) = \int_t^T P_1(\{r_1\}, t) h_1(s) ds = r_1^{\frac{1}{\eta}} \int_t^T y_1^{\frac{1}{\eta}}(s) h_1(s) ds, \tag{10.4}$$

where $h_1(s)$ is the probability density of waiting time until the first arrival of a request with a required response rate of greater than $y_1(s)$. The objective of the CSP is to determine the optimal allocation threshold $y_1^*(t)$ such that the expected revenue is maximized, i.e.

$$y_1^*(t) = \arg\max r_1^{\frac{1}{\eta}} \int_t^T y_1^{\frac{1}{\eta}}(s) h_1(s) ds. \tag{10.5}$$

The optimal dynamic threshold for the case of a single VMI can be determined using the following theorem.

Theorem 10.1   *If the CSP has a single available VMI at time t, then it is optimal to allocate it to a requesting application if the required delay tolerance $x_i \geq y_1^*(t)$, where the optimal threshold $y_1^*(t)$ solves the following equation:*

$$y_1^*(t) = \left(\frac{1 - F_{\hat{X}}((y_1^*)^{\frac{1}{\eta}}(t))}{f_{\hat{X}}((y_1^*)^{\frac{1}{\eta}}(t))} + \lambda \int_t^T \frac{(1 - F_{\hat{X}}((y_1^*)^{\frac{1}{\eta}}(s)))^2}{f_{\hat{X}}((y_1^*)^{\frac{1}{\eta}}(s))} ds \right)^{\eta}. \tag{10.6}$$

Proof: The proof for the dynamically optimal revenue maximizing curves requires the characterization of the expected revenue of the CSP over the allocation period is a dynamic threshold is used to filter incoming requests. Consider the case where at time t only a single VMI with response rate r_1 is available to the CSP for allocation up to time T. The expected revenue with a single cutoff curve $y_1(t)$ can be expressed as follows:

$$\mathcal{R}(\{r_1\}, t) = r_1^{\frac{1}{\eta}} \int_t^T y_1^{\frac{1}{\eta}}(s) h_1(s) ds, \tag{10.7}$$

where $h_1(s)$ is the probability density of waiting time until the first arrival of a request with a required response rate of greater than $y_1(s)$. The density can be represented by the first arrival in a non-homogeneous Poisson process with intensity $\lambda(1 - F_X(y_1(s)))$. For the sake of analytical tractability, we let $\hat{X} = X^{\frac{1}{\eta}}$. Therefore we will use the intensity $\lambda(1 - F_{\hat{X}}(y_1^{\frac{1}{\eta}}(s)))$ instead of the former one. Note that the density of homogeneous Poisson task arrival process is thinned by a factor $(1 - F_{\hat{X}}(y_1^{\frac{1}{\eta}}(s)))$, which represents the probability that the performance improvement of the arriving task is above the set qualification threshold $y_1^{\frac{1}{\eta}}(s)$. The density can be expressed as follows [114]:

$$h_1(s) = \lambda(1 - F_{\hat{X}}(y_1^{\frac{1}{\eta}}(s))) \exp\left(-\int_t^s \lambda(1 - F_{\hat{X}}(y_1^{\frac{1}{\eta}}(u)))du\right), \quad t \leq s \leq T. \quad (10.8)$$

Let $H(s) = \int_t^s \lambda\left(1 - F_{\hat{X}}(y_1^{\frac{1}{\eta}}(u))\right)du$. The expected revenue can then be written as follows:

$$\mathcal{R}(\{r_1\}, t) = r_1^{\frac{1}{\eta}} \int_t^T F_{\hat{X}}^{-1}\left(1 - \frac{H'(s)}{\lambda}\right) H'(s)e^{-H(s)}ds. \quad (10.9)$$

The kernel of integration can be expressed as follows:

$$L(s, H(s), H'(s)) = F_{\hat{X}}^{-1}\left(1 - \frac{H'(s)}{\lambda}\right) H'(s)e^{-H(s)}. \quad (10.10)$$

In order to maximize the expected revenue in (10.9), we employ the Euler–Lagrange equation from the calculus of variations [50]. The necessary condition for the revenue maximizing cutoff curves can be expressed as

$$\frac{\partial L(s, H(s), H'(s))}{\partial H(s)} - \frac{d}{dt}\frac{\partial L(s, H(s), H'(s))}{\partial H'(s)} = 0. \quad (10.11)$$

The partial derivatives can be expressed as follows:

$$\frac{\partial L}{\partial H(s)} = -F_{\hat{X}}^{-1}\left(1 - \frac{H'(s)}{\lambda}\right) H'(s)e^{-H(s)}, \quad (10.12)$$

$$\frac{\partial L}{\partial H'(s)} = e^{-H(s)}\left(F_{\hat{X}}^{-1}\left(1 - \frac{H'(s)}{\lambda}\right) - \left(\frac{H'(s)}{\lambda F_{\hat{X}}\left(F_{\hat{X}}^{-1}\left(1 - \frac{H'(s)}{\lambda}\right)\right)}\right)\right), \quad (10.13)$$

$$\frac{d}{dt}\frac{\partial L}{\partial H'(s)} = -\left(F_{\hat{X}}^{-1}\left(1 - \frac{H'(s)}{\lambda}\right) - \left(\frac{H'(s)}{\lambda F_{\hat{X}}\left(F_{\hat{X}}^{-1}\left(1 - \frac{H'(s)}{\lambda}\right)\right)}\right)\right)e^{-H(s)}H'(s)$$

$$+ e^{-H(s)}\left(\frac{-2H''(s)}{\lambda F_{\hat{X}}\left(F_{\hat{X}}^{-1}\left(1 - \frac{H'(s)}{\lambda}\right)\right)}\right.$$

$$\left. - \frac{F'_{\hat{X}}\left(F_{\hat{X}}^{-1}\left(1 - \frac{H'(s)}{\lambda}\right)\right)H'(s)H''(s)}{\lambda^2 F_{\hat{X}}^3\left(F_{\hat{X}}^{-1}\left(1 - \frac{H'(s)}{\lambda}\right)\right)}\right). \tag{10.14}$$

The Euler–Lagrange equation can be written as follows:

$$-(H'(s))^2 + 2H''(s) + \frac{F'_{\hat{X}}\left(F_{\hat{X}}^{-1}\left(1 - \frac{H'(s)}{\lambda}\right)\right)H'(s)H''(s)}{\lambda f_{\hat{X}}^2\left(F_{\hat{X}}^{-1}\left(1 - \frac{H'(s)}{\lambda}\right)\right)} = 0. \tag{10.15}$$

Then, plugging back $H(s) = \int_t^s \lambda(1 - F_{\hat{X}}(y_1^{\frac{1}{\eta}}(u)))du$, we obtain the following:

$$-\lambda(1 - F_{\hat{X}}(y_1^{\frac{1}{\eta}}(s)))^2 - \frac{2}{\eta}F_{\hat{X}}(y_1^{\frac{1}{\eta}}(s))y_1^{\frac{1}{\eta}-1}(s)y_1'(s)$$

$$- \frac{F'_{\hat{X}}(y_1^{\frac{1}{\eta}}(s))(1 - F_{\hat{X}}(y_1^{\frac{1}{\eta}}(s)))y_1^{\frac{1}{\eta}-1}(s)y_1'(s)}{\eta f_{\hat{X}}(y_1(s))} = 0. \tag{10.16}$$

This can be further expressed as follows:

$$\frac{d}{ds}(y_1^{\frac{1}{\eta}}(s)) + \frac{\lambda(1 - F_{\hat{X}}(y_1^{\frac{1}{\eta}}(s)))^2}{F_{\hat{X}}(y_1^{\frac{1}{\eta}}(s))} = \frac{d}{ds}\left(\frac{1 - F_{\hat{X}}(y_1(s))}{f_{\hat{X}}(y_1(s))}\right). \tag{10.17}$$

Integrating both sides from t to T results in the following:

$$y_1^{\frac{1}{\eta}}(T) - y_1^{\frac{1}{\eta}}(t) + \lambda \int_t^T \frac{(1 - F_{\hat{X}}(y_1^{\frac{1}{\eta}}(s)))^2}{f_{\hat{X}}(y_1^{\frac{1}{\eta}}(s))}ds = \left(\frac{1 - F_{\hat{X}}(y_1^{\frac{1}{\eta}}(T))}{f_{\hat{X}}(y_1^{\frac{1}{\eta}}(T))}\right)$$

$$- \left(\frac{1 - F_{\hat{X}}(y_1^{\frac{1}{\eta}}(t))}{f_{\hat{X}}(y_1^{\frac{1}{\eta}}(t))}\right). \tag{10.18}$$

Using the boundary condition $y_1^{\frac{1}{\eta}}(T) - \frac{1 - F_{\hat{X}}(y_1^{\frac{1}{\eta}}(T))}{f_{\hat{X}}(y_1^{\frac{1}{\eta}}(T))} = 0$, we reach the following expression:

$$-y_1^{\frac{1}{\eta}}(t) + \lambda \int_t^T \frac{(1 - F_{\hat{X}}(y_1^{\frac{1}{\eta}}(s)))^2}{f_{\hat{X}}(y_1^{\frac{1}{\eta}}(s))}ds = -\left(\frac{1 - F_{\hat{X}}(y_1^{\frac{1}{\eta}}(t))}{f_{\hat{X}}(y_1^{\frac{1}{\eta}}(t))}\right). \tag{10.19}$$

Rearranging the terms results in the following:

$$y_1^{\frac{1}{n}}(t) = \frac{1 - F_{\hat{X}}(y_1^{\frac{1}{n}}(t))}{f_{\hat{X}}(y_1^{\frac{1}{n}}(t))} + \lambda \int_t^T \frac{(1 - F_{\hat{X}}(y_1^{\frac{1}{n}}(s)))^2}{f_{\hat{X}}(y_1^{\frac{1}{n}}(s))} ds. \tag{10.20}$$

Equivalently, it can be written as follows:

$$y_1(t) = \left(\frac{1 - F_{\hat{X}}(y_1^{\frac{1}{n}}(t))}{f_{\hat{X}}(y_1^{\frac{1}{n}}(t))} + \lambda \int_t^T \frac{(1 - F_{\hat{X}}(y_1^{\frac{1}{n}}(s)))^2}{f_{\hat{X}}(y_1^{\frac{1}{n}}(s))} ds \right)^n. \tag{10.21}$$

To complete the proof, we note that the expected revenue is given by $R(r_j, t) = r_j^{\frac{1}{n}} R(1, t)$ where

$$R(1, t) = \int_t^T y_1^{\frac{1}{n}}(s) \lambda (1 - F_{\hat{X}}(y_1^{\frac{1}{n}}(s))) e^{-\int_t^s \lambda(1 - F_{\hat{X}}(y_1^{\frac{1}{n}}(s))) dz} ds. \tag{10.22}$$

Differentiating the earlier expression with respect to t gives the following:

$$R'(1, t) = \lambda (1 - F_{\hat{X}}(y_1^{\frac{1}{n}}(t)))(R(1, t) - y_1^{\frac{1}{n}}(t)). \tag{10.23}$$

It can be shown that $R(1, t) = \lambda \int_t^T \frac{(1 - F_{\hat{X}}(y_1^{\frac{1}{n}}(s)))^2}{f_{\hat{X}}(y_1^{\frac{1}{n}}(s))} ds$ satisfies Eq. (10.23). $\qquad \square$

In the subsequent subsection, we extend the approach to the case of multiple available VMIs.

10.3.2 Multiple VMI Case

In this section, we first present the case of two available VMIs at the CSP and then generalize it to the case of multiple available VMIs. If there are only two available VMIs with response rates r_1 and r_2, then the expected revenue of the CSP can be expressed as follows:

$$\begin{aligned}
R(\{r_1, r_2\}, t) \\
&= \int_0^T (P_2(\{r_1, r_2\}t) + R(\{r_1\}, t)) h_2(t) dt + \int_0^T (P_1(\{r_1, r_2\}t) \\
&\quad + R(\{r_2\}, t)) h_1(t) dt, \\
&= \int_0^T \left(r_2^{\frac{1}{n}} y_2^{\frac{1}{n}} + R(r_1, t) \right) \lambda (1 - F_{\hat{X}}(y_2^{\frac{1}{n}}(t))) e^{-\int_0^t \lambda(1 - F_{\hat{X}}(y_2^{\frac{1}{n}}(s))) ds} dt \\
&\quad + (r_1^{\frac{1}{n}} - r_2^{\frac{1}{n}}) \int_0^T \left(y_1^{\frac{1}{n}}(t) - R(1, t) \right) \lambda (1 - F_{\hat{X}}(y_1^{\frac{1}{n}}(t))) e^{-\int_0^t \lambda(1 - F_{\hat{X}}(y_2^{\frac{1}{n}}(s))) ds} dt,
\end{aligned} \tag{10.24}$$

where $h_1(t)$ represents the density of waiting time till the first arrival of a request with a required response rate of atleast $y_1(t)$ if no request with a required response rate in the interval $[y_2(t), y_1(t))$ has arrived. Similarly, $h_2(t)$ represents the density of waiting time till the first arrival of a request with a required response rate in the interval $[y_2(t), y_1(t))$ if no request with a required response rate in the interval $[y_1(t), \infty)$ has arrived. The optimization problem in this case becomes the following.

$$y_2^*(t) = \arg\max \int_0^T \left(r_2^{\frac{1}{\eta}} (y_2^*)^{\frac{1}{\eta}} + R(r_1, t) \right) \lambda \left(1 - F_{\hat{X}} \left((y_2^*)^{\frac{1}{\eta}}(t) \right) \right) e^{-\int_0^t \lambda(1 - F_{\hat{X}}((y_2^*)^{\frac{1}{\eta}}(s)))ds} dt$$

$$+ \left(r_1^{\frac{1}{\eta}} - r_2^{\frac{1}{\eta}} \right) \int_0^T \left((y_1^*)^{\frac{1}{\eta}}(t) - R(1, t) \right) \lambda \left(1 - F_{\hat{X}} \left((y_1^*)^{\frac{1}{\eta}}(t) \right) \right) e^{-\int_0^t \lambda(1 - F_{\hat{X}}((y_2^*)^{\frac{1}{\eta}}(s)))ds} dt.$$

(10.25)

The optimal dynamic threshold for the case of two available VMIs can be determined using the following theorem.

Theorem 10.2 *If the CSP has a two available VMIs at time t, then it is optimal to allocate the low response rate VMI to a requesting application if the required delay tolerance $x_i \in [y_2^*(t), y_1^*(t))$, and to allocate the high response rate VMI if the required delay tolerance $x_i \geq y_2^*(t)$ where the optimal threshold $y_2^*(t)$ solves the following equation:*

$$y_2^*(t) = \left(\frac{1 - F_{\hat{X}}((y_2^*)^{\frac{1}{\eta}}(t))}{F_{\hat{X}}((y_2^*)^{\frac{1}{\eta}}(t))} + \lambda \int_t^T \frac{(1 - F_{\hat{X}}((y_2^*)^{\frac{1}{\eta}}(s)))^2}{F_{\hat{X}}((y_2^*)^{\frac{1}{\eta}}(s))} ds - R(1, t) \right)^{\eta}.$$

(10.26)

Proof: If two VMIs are available, then the revenue can be expressed as follows:

$$\int_0^T (P_2(\{r_1, r_2\}, t) + R(\{r_1\}, t))h_2(t)dt + \int_0^T (P_1(\{r_1, r_2\}, t) + R(\{r_2\}, t))h_1(t)dt.$$

(10.27)

This can be further written as follows:

$$\int_0^T \left(r_2^{\frac{1}{n}} y_2^{\frac{1}{n}} + R(r_1, t) \right) \lambda(1 - F_{\hat{X}}(y_2^{\frac{1}{n}}(t))) e^{-\int_0^t \lambda(1 - F_{\hat{X}}(y_2^{\frac{1}{n}}(s)))ds} dt$$

$$+ \int_0^T \left((r_1^{\frac{1}{n}} - r_2^{\frac{1}{n}}) y_1^{\frac{1}{n}}(t) + R(r_2, t) - R(r_1, t) \right)$$

$$\times \lambda(1 - F_{\hat{X}}(y_1^{\frac{1}{n}}(t))) e^{-\int_0^t \lambda(1 - F_{\hat{X}}(y_2^{\frac{1}{n}}(s)))ds} dt,$$

$$= \int_0^T \left(r_2^{\frac{1}{n}} y_2^{\frac{1}{n}} + R(r_1, t) \right) \lambda(1 - F_{\hat{X}}(y_2^{\frac{1}{n}}(t))) e^{-\int_0^t \lambda(1 - F_{\hat{X}}(y_2^{\frac{1}{n}}(s)))ds} dt$$

$$+ (r_1^{\frac{1}{n}} - r_2^{\frac{1}{n}}) \int_0^T \left(y_1^{\frac{1}{n}}(t) - R(1, t) \right) \times \lambda(1 - F_{\hat{X}}(y_1^{\frac{1}{n}}(t))) e^{-\int_0^t \lambda(1 - F_{\hat{X}}(y_2^{\frac{1}{n}}(s)))ds} dt.$$

$$(10.28)$$

Let $G(t) = \int_0^t \lambda(1 - F_{\hat{X}}(y_1^{\frac{1}{n}}(s)))ds$ and $H(t) = \int_0^t \lambda(1 - F_{\hat{X}}(y_2^{\frac{1}{n}}(s)))ds$. Then, the expression can further be written as follows:

$$\int_0^T \left(r_2^{\frac{1}{n}} F_{\hat{X}}^{-1} \left(1 - \frac{H'(t)}{\lambda} \right) + r_1^{\frac{1}{n}} R(1, t) \right) H'(t) e^{-H(t)} dt$$

$$+ (r_1^{\frac{1}{n}} - r_2^{\frac{1}{n}}) \int_0^T \left(F_{\hat{X}}^{-1} \left(1 - \frac{G'(t)}{\lambda} \right) \right) G''(t) e^{-H(t)} dt. \qquad (10.29)$$

Therefore,

$$L_1(t, H(t), H'(t)) = \left(r_2^{\frac{1}{n}} F_{\hat{X}}^{-1} \left(1 - \frac{H'(t)}{\lambda} \right) + r_1^{\frac{1}{n}} R(r_1, t) \right) H'(t) e^{-H(t)}$$

$$+ (r_1^{\frac{1}{n}} - r_2^{\frac{1}{n}}) \left(F_{\hat{X}}^{-1} \left(1 - \frac{G'(t)}{\lambda} \right) - R(1, t) \right) G'(t) e^{-H(t)}.$$

Computing the Euler–Lagrange equation, $\frac{\partial L_1}{\partial H(t)} - \frac{d}{dt} \frac{\partial L_1}{\partial H'(t)} = 0$ results in the following:

$$- (r_1^{\frac{1}{n}} - r_2^{\frac{1}{n}}) G'(t) \left(F_{\hat{X}}^{-1} \left(1 - \frac{G'(t)}{\lambda} \right) - R(1, t) \right)$$

$$- r_1^{\frac{1}{n}} R'(1, t) - r_2^{\frac{1}{n}} \frac{(H'(t))^2}{\lambda F_{\hat{X}} \left(F_{\hat{X}}^{-1} \left(1 - \frac{H'(t)}{\lambda} \right) \right)} + 2 r_2^{\frac{1}{n}} \frac{H''(t)}{\lambda F_{\hat{X}} \left(F_{\hat{X}}^{-1} \left(1 - \frac{H'(t)}{\lambda} \right) \right)}$$

$$+ r_2^{\frac{1}{n}} \frac{F_{\hat{X}}' \left(F_{\hat{X}}^{-1} \left(1 - \frac{H'(t)}{\lambda} \right) \right) H'(t) H''(t)}{\lambda^2 \left(F_{\hat{X}} \left(F_{\hat{X}}^{-1} \left(1 - \frac{H'(t)}{\lambda} \right) \right) \right)^3} = 0, \qquad (10.30)$$

and $\frac{\partial L_1}{\partial G(t)} - \frac{d}{dt}\frac{\partial L_1}{\partial G'(t)} = 0$ results in the following:

$$
-H'(t)\left(-\frac{G'(t)}{\lambda F_{\hat{X}}\left(F_{\hat{X}}^{-1}\left(1 - \frac{G'(t)}{\lambda}\right)\right)} + F_{\hat{X}}^{-1}\left(1 - \frac{G'(t)}{\lambda}\right) - R(1,t)\right)
$$

$$
-R'(1,t) - \frac{2G''(t)}{\lambda F_{\hat{X}}\left(F_{\hat{X}}^{-1}\left(1 - \frac{G'(t)}{\lambda}\right)\right)} + \frac{f_{\hat{X}}'\left(F_{\hat{X}}^{-1}\left(1 - \frac{G'(t)}{\lambda}\right)\right)G'(t)G''(t)}{\lambda^2 f_{\hat{X}}^3\left(F_{\hat{X}}^{-1}\left(1 - \frac{G'(t)}{\lambda}\right)\right)} = 0.
$$

(10.31)

Eventually, it leads to the following differential equations:

$$
-(r_1^{\frac{1}{\eta}} - r_2^{\frac{1}{\eta}})\lambda\left(1 - F_{\hat{X}}\left(y_1^{\frac{1}{\eta}}(t)\right)\right)(y_1^{\frac{1}{\eta}}(t) - R(1,t))
$$

$$
-r_1^{\frac{1}{\eta}}(t)R'(1,t) - r_2^{\frac{1}{\eta}}\frac{\lambda(1 - F_{\hat{X}}(y_2^{\frac{1}{\eta}}(t)))^2}{F_{\hat{X}}(y_2^{\frac{1}{\eta}}(t))} - 2r_2^{\frac{1}{\eta}}\frac{y_2'(t)y_2^{\frac{1}{\eta}-1}(t)}{\eta}
$$

$$
-r_2^{\frac{1}{\eta}}\frac{y_2'(t)(1 - F_{\hat{X}}(y_2^{\frac{1}{\eta}}(t)))F_{\hat{X}}'(y_2^{\frac{1}{\eta}}(t))y_2^{\frac{1}{\eta}-1}(t)}{\eta F_{\hat{X}}^2(y_2^{\frac{1}{\eta}}(t))} = 0,
$$

(10.32)

and

$$
\frac{2y_1'(t)y_1^{\frac{1}{\eta}-1}(t)}{\eta} - R'(1,t) - \frac{F_{\hat{X}}'(y_1^{\frac{1}{\eta}}(t))(1 - F_{\hat{X}}(y_1^{\frac{1}{\eta}}(t)))y_1'(t)y_1^{\frac{1}{\eta}-1}(t)}{\eta f_{\hat{X}}^2(y_1^{\frac{1}{\eta}}(t))}
$$

$$
+ \lambda(1 - F_{\hat{X}}(y_2^{\frac{1}{\eta}}(t)))\left(\frac{(1 - F_{\hat{X}}(y_1^{\frac{1}{\eta}}(t)))}{f_{\hat{X}}(y_1^{\frac{1}{\eta}}(t))} + y_1^{\frac{1}{\eta}}(t) + R(1,t)\right) = 0. \quad (10.33)
$$

Now, we show that a solution to these differential equations is given by the solution to the following system of equations:

$$
y_1(t) = \left(\frac{1 - F_{\hat{X}}(y_1^{\frac{1}{\eta}}(t))}{f_{\hat{X}}(y_1^{\frac{1}{\eta}}(t))} + \lambda\int_t^T\frac{(1 - F_{\hat{X}}(y_1^{\frac{1}{\eta}}(s)))^2}{f_{\hat{X}}(y_1^{\frac{1}{\eta}}(s))}ds\right)^\eta,
$$

(10.34)

and

$$
y_2(t) = \left(\frac{1 - F_{\hat{X}}(y_2^{\frac{1}{\eta}}(t))}{f_{\hat{X}}(y_2^{\frac{1}{\eta}}(t))} + \lambda\int_t^T\frac{(1 - F_{\hat{X}}(y_2^{\frac{1}{\eta}}(s)))^2}{f_{\hat{X}}(y_2^{\frac{1}{\eta}}(s))}ds - R(1,t)\right)^\eta.
$$

(10.35)

Differentiating (10.34) with respect to t results in the following:

$$2\frac{y_1''^{\frac{1}{\eta}-1}(t)y_1'(t)}{\eta} = -y_1'(t)\frac{(1 - F_{\hat{X}}(y_1''^{\frac{1}{\eta}}(t)))f'(y_1''^{\frac{1}{\eta}}(t))}{f_{\hat{X}}^2(y_1''^{\frac{1}{\eta}}(t))} - \frac{\lambda(1 - F_{\hat{X}}(y_1''^{\frac{1}{\eta}}(t)))^2}{f_{\hat{X}}(y_1''^{\frac{1}{\eta}}(t))}.$$

(10.36)

Substituting this in (10.33) results in the following:

$$\left(\lambda(1 - F_{\hat{X}}(y_1''^{\frac{1}{\eta}}(t))) - \lambda(1 - F_{\hat{X}}(y_2''^{\frac{1}{\eta}}(t)))\right) \times \left(\frac{1 - F_{\hat{X}}(y_1''^{\frac{1}{\eta}}(t))}{f_{\hat{X}}(y_1''^{\frac{1}{\eta}}(t))} - y_1''(t) + R(1,t)\right) = 0,$$

(10.37)

which is satisfied for all $y_2(t)$ using (10.34). Differentiating (10.35) with respect to t results in the following:

$$2\frac{y_2'(t)y_2''^{\frac{1}{\eta}-1}}{\eta} = -\frac{(1 - F_{\hat{X}}(y_2(t))F_{\hat{X}}'(y_2(t))y_2'(t)}{f_{\hat{X}}^2(y_2(t))} - \frac{\lambda(1 - F_{\hat{X}}(y_2(t)))^2}{F_{\hat{X}}(y_2(t))} - R'(1,t).$$

(10.38)

Substituting this into (10.33) results in the following:

$$-(r_1''^{\frac{1}{\eta}} - r_2''^{\frac{1}{\eta}})\lambda(1 - F_{\hat{X}}(y_1''^{\frac{1}{\eta}}(t)))(y_1''^{\frac{1}{\eta}(t)} - R(1,t)) - R'(1,t)\left(r_1''^{\frac{1}{\eta}} - r_2''^{\frac{1}{\eta}}\right) = 0. \quad (10.39)$$

This holds for $y_1(t)$ from (10.23). $\qquad\square$

Note that the optimal threshold $y_2^*(t)$ relies on obtaining the threshold $y_1^*(t)$. In the general case, it can be shown by induction that the optimal thresholds can be obtained recursively using the following theorem:

Theorem 10.3 *If there are N_t available VMIs at time t, then it is optimal to allocated a VMI with response rate r_i to an incoming request with minimum required response rate $x_i \in [y_i^*(t), y_{i-1}^*(t)]$, where the optimal dynamic thresholds $y_i^*(t)$ satisfy the following recursive equation:*

$$y_i^*(t) = \left(\frac{1 - F_{\hat{X}}(((y_i^*)^{\frac{1}{\eta}})(t))}{f_X(((y_i^*)^{\frac{1}{\eta}})(t))} + \lambda\int_t^T \frac{(1 - F_{\hat{X}}(((y_{i-1}^*)^{\frac{1}{\eta}})(s)))^2}{f_X(((y_{i-1}^*)^{\frac{1}{\eta}})(s))}ds\right.$$

$$\left. -\lambda\int_t^T \frac{(1 - F_{\hat{X}}(((y_i^*)^{\frac{1}{\eta}})(s)))^2}{f_X(((y_i^*)^{\frac{1}{\eta}})(s))}ds\right)^{\eta}, \quad i = 2,\ldots,N_t. \quad (10.40)$$

Proof: To prove the general case, we show that when two VMIs with unit characteristic are available, the revenue can be computed as follows:

$$R(\{1,1\},t) = \int_t^T \left(y_2^{\frac{1}{\eta}}(t) + R(1,s) \right) \lambda(1 - F_{\hat{X}}(y_2^{\frac{1}{\eta}}(s)))e^{-\int_t^s \lambda(1-F_{\hat{X}}(y_2^{\frac{1}{\eta}}(z)))dz} ds.$$

(10.41)

Differentiating with respect to t results in

$$R'(\{1,1\},t) = \lambda(1 - F_{\hat{X}}(y_2^{\frac{1}{\eta}}(t)))(R(\{1,1\}) - y_2^{\frac{1}{\eta}}(t) - R(1,t)),$$

(10.42)

It can be shown that $R(\{1,1\},t) = \int_t^T \frac{(1-F_{\hat{X}}(y_2^{\frac{1}{\eta}}(s))^2)}{f_{\hat{X}}(y_2^{\frac{1}{\eta}}(s))} ds$ satisfies the aforementioned differential equation using (10.35). Using a similar procedure, it can be shown that in the general case, the optimal threshold solves the following equation:

$$y_i(t) = \left(\frac{1 - F_{\hat{X}}(y_i^{\frac{1}{\eta}}(t))}{f_{\hat{X}}(y_i^{\frac{1}{\eta}}(t))} + R(\mathbf{1}_i,t) - R(\mathbf{1}_{i-1},t) \right)^{\eta},$$

(10.43)

where

$$R(\mathbf{1}_i,t) - \lambda \int_t^T \frac{(1 - F_{\hat{X}}(y_i^{\frac{1}{\eta}}(s)))^2}{f_{\hat{X}}(y_i^{\frac{1}{\eta}}(s))} ds.$$

(10.44)

□

Note that the optimal cutoff curves are independent of the response rates of the available VMIs. In fact, they depend only on the number of available VMIs at time t and on the statistical information about the sequentially arriving computational requests by MC-IoT applications. In the following set of corollaries, we provide the results obtained for special cases of the statistical information about arriving requests.

As described in Section 10.2.1, the allocation of available VMIs is decided based on a partition of the users' performance improvement characteristic that are defined by cutoff curves $y_i(t), i = 1, \ldots, N_t$.

These dynamic cutoff curves act as a lookup table for the CSP to instantly decide the allocation of the available VMIs to MC-IoT requests. The optimal dynamic cutoff curves that maximize the expected revenue of the CSP, based on the QoE provided to the users, in the presence of uncertain sequentially arriving requests by MC-IoT applications can be obtained using the following theorem [52]:

Corollary 10.1 *Assuming that the transformed required response rate of sequentially arriving MC-IoT applications, denoted by $\hat{X}$ (see Proof of*

Theorem 10.1), follows an exponential distribution with a mean of α^{-1}, i.e. $f_{\hat{X}}(x) = \alpha e^{-\alpha x}$, and $F_{\hat{X}}(x) = 1 - e^{-\alpha x}$, then the optimal cutoff curves for allocation can be expressed as follows:

$$y_1^*(t) = \frac{1}{\alpha^\eta}\left[1 + \log\left(1 + \frac{\lambda(T-t)}{e}\right)\right]^\eta, \tag{10.45}$$

$$y_2^*(t) = \frac{1}{\alpha^\eta}\left[1 + \log\left(1 + \frac{\lambda^2(T-t)^2}{2e(\lambda(T-t)+e)}\right)\right]^\eta, \tag{10.46}$$

$$y_3^*(t) = \frac{1}{\alpha^\eta}\left[1 + \log\left(1 + \frac{\lambda^3(T-t)^3}{3e(\lambda^2(T-t)^2 + 2e(\lambda(T-t)+e))}\right)\right]^\eta. \tag{10.47}$$

The remaining lower cutoff curves cannot be easily derived in closed form and can be computed numerically.

Corollary 10.2 *Assuming that the transformed required response rate of sequentially arriving MC-IoT applications, denoted by $\hat{X}$ (see proof of Theorem 10.1), is uniformly distributed in the interval $[0, \beta]$, i.e. $f_{\hat{X}}(x) = \frac{1}{\beta}$ and $F_{\hat{X}}(x) = \frac{x-\beta}{\beta}, x \in [0, \beta]$, then the optimal cutoff curves for the allocation of the best available VMI can be expressed as follows:*

$$y_1^*(t) = \beta^\eta\left(1 - \frac{2}{\lambda(T-t)+4}\right)^\eta. \tag{10.48}$$

The lower cutoff curves cannot be easily obtained analytically and thus require numerical evaluation.

10.3.3 Expected Revenue

The expected revenue from the optimal allocation of VMIs to sequentially arriving MC-IoT requests can be expressed by the following theorem:

Theorem 10.4 *The expected revenue of the CSP at time t if a total of N_t VMIs with response rated defined by $\mathbf{r}_t = [r_1, r_2, \ldots, r_{N_t}]$ can be expressed as:*

$$\mathcal{R}(\mathbf{r}_t, t) = \sum_{i=1}^{N_t} r_i^{\frac{1}{\eta}}\left(y_i^{\frac{1}{\eta}}(t) - \frac{1 - F_{\hat{X}}(y_i^{\frac{1}{\eta}}(t))}{f_X(y_i^{\frac{1}{\eta}}(t))}\right). \tag{10.49}$$

Proof: The expected revenue if only a single VMI with unit response rate is available is given by

$$\mathcal{R}(\{1\}, t) = \int_t^s \frac{(1 - F_{\hat{X}}(y_1^{\frac{1}{\eta}}(s)))^2}{f_X(y_1^{\frac{1}{\eta}}(s))}ds = \left(y_1^{\frac{1}{\eta}}(t) - \frac{1 - F_{\hat{X}}(y_1^{\frac{1}{\eta}}(t))}{f_X(y_1^{\frac{1}{\eta}}(t))}\right).$$

Similarly, if two VMIs with unit response rates are available, then using Theorem 10.1, the expected revenue is given by

$$
\mathcal{R}(\{1,1\},t) = \int_t^s \frac{(1 - F_{\hat{X}}(y_2^{\frac{1}{n}}(s)))^2}{f_X(y_2^{\frac{1}{n}}(s))} ds
$$

$$
= \left(y_1^{\frac{1}{n}}(t) - \frac{1 - F_{\hat{X}}(y_1^{\frac{1}{n}}(t))}{f_X(y_1^{\frac{1}{n}}(t))} \right) + \left(y_2^{\frac{1}{n}}(t) - \frac{1 - F_{\hat{X}}(y_2^{\frac{1}{n}}(t))}{f_X(y_2^{\frac{1}{n}}(t))} \right).
$$

Using an inductive argument along with the fact that $\mathcal{R}(\{r_j\},t) = r_j^{\frac{1}{n}}\mathcal{R}(\{1\},t)$ proves the result. $\qquad\square$

Notice that the expected revenue is linear in the response rates of the available VMIs and increases if a high response rate or equivalently a low latency is provided by the VMIs at the fog nodes.

10.3.4 Implementation of Dynamic VMI Allocation and Pricing

In this section, we explain the operation of the proposed QoE based revenue maximizing dynamic allocation and pricing framework. Requests for remote computations by MC-IoT applications arrive at the CSP at random times and have a delay sensitivity, which is unknown *a priori*. Based on the initial number of available VMIs at the CSP N_0, there is a minimum cutoff threshold $y_{N_0}(t)$ that an incoming request has to cross before being allocated a VMI. If the incoming request qualifies for an allocation, the CSP needs to decide which VM to allocate to the user. A lookup table denoted by $\mathcal{T}$ is prepared by the CSP, which contains pre-computed optimal dynamic cutoff curves determined in Section 10.3.2. Using this lookup table, the request is optimally classified for allocation to one of the available VMIs. The data from the MC-IoT request is forwarded to the selected VMI and a price, also available in the lookup table, is charged to the requesting application. Once the allocation has been completed, the VMI is removed from the set of available VMIs. The lookup tables are updated by removing the least cutoff threshold $y_{N_0}(t)$. The remaining available VMIs are re-arranged in descending order and their prices are updated in the lookup table $\mathcal{T}$. This process is repeated until either all the available VMIs have been successfully allocated or the allocation period has ended. This procedure has been summarized in Algorithm 10.1 and the associated flow diagram is provided in Figure 10.1.

Algorithm 10.1 Dynamic VMI Allocation and Pricing

Require: Initialize counter $= 0$, request index $i = 1$, starting time $t = 0$, $\mathbf{r}_0 = \{r_i, i = 1, \ldots, N_0 : r_1 \geq r_2 \geq \ldots \geq r_{N_0}\}$.

1: **while** counter $< N_0$ **and** $t < T$ **do**

2:      Determine the required response rate of the arriving computational request x_i.

3:      **if** $x_i \geq y_{N_0-\text{counter}}(t)$ **then**

4:          Classify the request using the lookup table, i.e. determine $j : x_i \in [y_j(t), y_{j-1}(t)]$.

5:          The ith requesting application is allocated to the jth highest VMI.

6:          Use the mapping $\mathcal{M}$ to realize the allocation in terms of the fog node and available VMI.

7:          Use $\mathcal{T}$ to charge a price $P_j(\mathbf{r}_t)$ to the requesting application.

8:          Remove the VMI corresponding to r_j from the list of available VMIs.

9:          Re-arrange the set of available VMIs in descending order.

10:         Update lookup table for relevant cutoff curves $\mathcal{T} = \mathcal{T} \backslash y_{N_0-\text{counter}}(t)$.

11:         Update $\mathcal{T}$ with new prices corresponding to the updated list of available VMIs.

12:      **end if** $i \leftarrow i + 1$.

13: **end while**

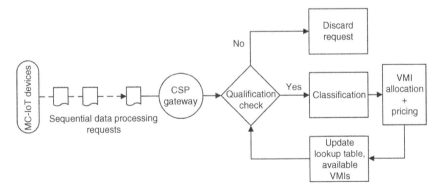

Figure 10.1 Resource allocation and pricing flow diagram.

10.4 Numerical Experiments and Discussion

In this section, we present the results of the numerical experiments performed to evaluate the performance of our proposed allocation and pricing framework. We first describe the simulation setup and parameter selection. We then provide the results of the simulation and study the behavior of the framework under varying parameters. Finally, we provide a comparison of the proposed framework with some benchmark allocation strategies and discuss the insights.

10.4.1 Experiment Setup

We assume a CSP with $k = 5$ fog data centers in addition to the main cloud servers (Internet data centers) serving a fixed MC-IoT installation area. Each fog data center has $n_i = 20, i = 1, \ldots, 5$, available VMIs for allocation to incoming MC-IoT requests. We have selected a small fog computing architecture in the experiments for the sake of simplicity. However, the experiments can be extended to larger topologies without loss of generality in the results.

The fog data centers are located at increasing distances from the MC-IoT devices, which result in increasing latency experienced by the devices while accessing the corresponding data centers. We assume the round trip time (RTT) of the fog nodes experienced by the MC-IoT installation are $l_1 = 0.1$, $l_2 = 0.2, l_3 = 0.4, l_4 = 0.6$, and $l_5 = 0.8$ ms, respectively. The transmission delay over the air interfaces is assumed to be fixed at $\tau^{(o)} = 0.1$ ms. The workload processing times of the VMIs, which depends on the provisioned computing resources, is considered to be uniformly distributed in the interval $[0.2, 1]$ ms in the simulations. A demand horizon of $T = 12$ hours is assumed during which the allocation takes place. The constant η in the QoE function is selected to be $\eta = 1$ for simplicity of results.

The MC-IoT applications generate computational requests according to a homogeneous Poisson process that arrive sequentially at the CSP with an arrival rate $\lambda = 10$ requests per hour unless otherwise specified. The reported required response rate of MC-IoT applications is simulated as i.i.d. random variables distributed according to an exponential distribution with mean α^{-1}, where $\alpha = 1$, or according to a uniform distribution in the interval $[0, \beta]$, where $\beta = 10$. In the following subsections, we provide the results of the simulations and the obtained insights.

10.4.2 Simulation Results

First we obtain the optimal dynamic cutoff curves obtained for dynamic revenue maximizing VMI allocation. Based on the assumption that the reported required response rate of MC-IoT applications is exponentially distributed and uniformly distributed, the cutoff curves are computed by iteratively solving the differential equations provided in Theorem 10.3. The first 20 dynamic thresholds for both cases are provided in Figure 10.2. These are used as a

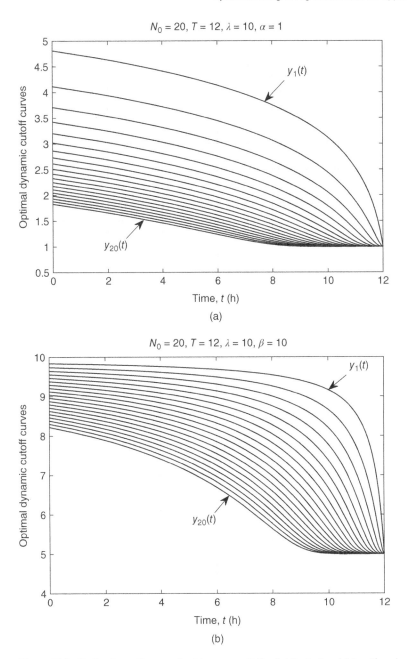

Figure 10.2 Optimal cutoff curves for (a) exponentially distributed and (b) uniformly distributed arrival characteristics.

policy for classification of the arriving requests. In general, it can be observed that the thresholds decrease, i.e. the qualification standard is reduced, as the time increases. This implies that the CSP becomes less and less selective as the allocation period is ending to ensure that the maximum number of VMIs are allocated to upcoming requests. However, when there is a lot of time remaining before the allocation period ends, the CSP is more selective ensuring that only the task that reports a sufficiently high required response rate is allocated to a VMI. It can also be observed from Figure 10.2a that the cutoff curves for the exponentially distributed arrivals have a decreasing allocation margin as the number of VMIs increase. This is due to the possibility of arrival of an extremely delay sensitive application for which the barrier is set extremely high. On the other hand, Figure 10.2b shows that under uniformly distributed arrival characteristic, the allocation margin for each of the VMIs is almost constant as the arrival characteristics are equally likely. In both cases, as time approaches the deadline, the cutoff curves approach a constant value and the allocation mechanism turns into a first-price auction mechanism.

Next, we investigate the behavior of the proposed allocation and pricing scheme in response to the changing rate of arrival of MC-IoT requests and the mean arrival characteristic for both exponentially and uniformly distributed arrival types. Figure 10.3 shows the expected revenue of the CSP and the expected QoE of the users as the rate of arrival or requests increases. It can be observed from Figure 10.3a,c that the expected revenue of the CSP increases as the arrival rate increases but saturates at high arrival rates as the available VMIs are exhausted. The exponential arrival characteristic results in a higher expected revenue in general due to the possibility of arrival of highly demanding and highly paying requests, which is not the case in the uniform case. Figure 10.3b,d depict a similar behavior in the expected QoE of the users in the two cases. Figure 10.4 investigates the behavior of the expected revenue of the CSP and the expected QoE of the users in response to a change in the mean of the arrival characteristic. It is observed from Figure 10.4a,c that the expected revenue increases linearly with the mean of the arrival characteristic for both exponential and uniform arrival types. This is because a higher arrival type raises the qualification standard for allocation and consequently the prices. Hence, it does not saturate as the arrival type or rate increases. Finally, Figure 10.4b,d shows a similar increasing behavior in the expected QoE of the users with an increase in the mean arrival characteristic.

10.4.3 Comparison with Other Approaches

To illustrate the performance of the developed optimal allocation mechanism in terms of the QoE of MC-IoT applications, we compare our proposed allocation scheme with the following benchmark strategies:

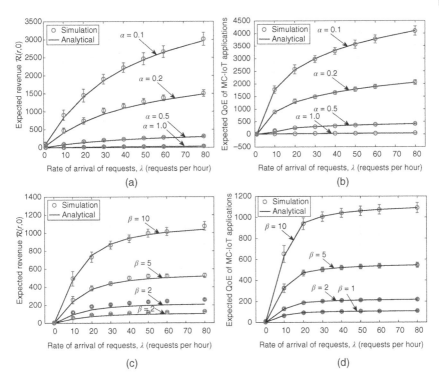

Figure 10.3 Effect of arrival rate on the expected revenue of the CSP (a, c) and expected QoE of the users (b,d) for an exponentially and uniformly distributed arrival characteristic respectively.

1. **Greedy Allocation**: In this strategy, the CSP uses a pessimistic approach toward future arrivals, i.e. assumes that more delay tolerant requests will arrive in the future as compared with the current one, and allocates the best available VMIs·to the requesting applications first. This strategy does not employ any foresight and only makes myopic decisions based on the current arriving task. Hence it attempts to maximize the QoE and the generated revenue by allocating the best available resources to incoming tasks first.

2. **Optimistic Allocation**: The optimistic allocation policy is the opposite of the greedy allocation policy. It always assumes that the arrivals in the future will be of more QoE sensitive that the current task at hand. Hence, it saves the best available VMIs while allocating the VMI with the highest end-to-end delay to the current request expecting more QoE sensitive requests to arrive in the future.

3. **Random Allocation**: In this strategy, the CSP randomly allocates one of the available VMI in any of the fog nodes to a requesting application. In other words, it has a uniform expectation about the nature of the requests that will

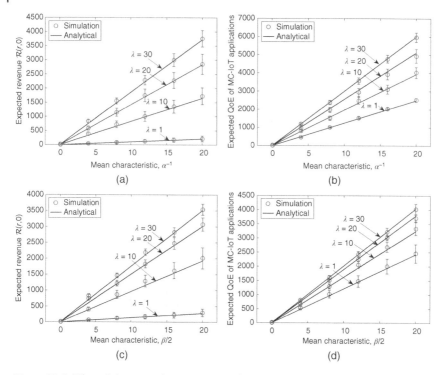

Figure 10.4 Effect of the mean characteristic on the expected revenue of the CSP (a, c) and expected QoE of the users (b, d) for an exponentially and uniformly distributed arrival characteristic respectively.

arrive in the future and hence allocates the available VMIs to an upcoming request with equal probability.

Finally, the proposed optimal dynamic strategy makes used of the statistical information about the future arriving tasks to make dynamically optimal decisions for allocating VMIs to arriving requests strategically. For the scenario considered in the simulations, we compare the proposed approach with each of the aforementioned approaches in terms of the average QoE experienced by the MC-IoT applications. The results of the expected QoE achieved by the users in comparison with the benchmark schemes are provided in Figure 10.5. Figure 10.5a,b illustrate the average QoE against varying the rate and mean characteristic of the respective arrival for an exponentially distributed required response rate. Figure 10.5c,d illustrate the average QoE against varying the rate and mean characteristic of the respective arrival for a uniformly distributed required response rate. It is clear from the results that the proposed allocation strategy provides the best expected QoE to the users as well as maximizes the revenue of the CSP. Furthermore, random pairings result in a very low expected

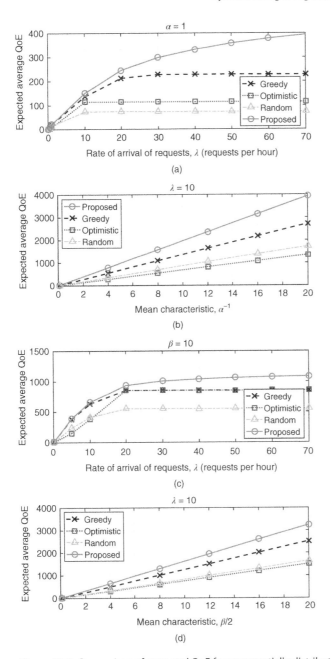

Figure 10.5 Comparison of expected QoE for exponentially distributed (a, b) and uniformly distributed (c, d) arrival characteristic.

QoE while the greedy and optimistic approaches lead to an intermediate average QoE of the users.

10.5 Summary and Conclusion

In this chapter, we provide a QoE based revenue maximizing dynamic resource allocation and pricing framework for fog-enabled MC-IoT applications. We propose an implementable mechanism for allocation of different VMIs available at the fog nodes that result in varying end-to-end delay for user applications. As opposed to existing works in the literature focusing on static assignment of computational tasks and available fog resources, the proposed resource allocation and pricing strategy is dynamic with instantaneous decision-making as well as takes the hierarchical fog-cloud architecture into account. The developed framework provides an optimal threshold based classification mechanism that uses statistical information about the MC-IoT requests arriving in the future to make dynamically optimal real-time allocation decisions resulting in maximizing revenue of the CSP, based on the QoE of the users. The dynamic thresholds can be pre-computed and used as a lookup table for real-time decision-making. Numerical results confirm that the proposed allocation scheme significantly performs better in terms of the QoE achieved by the users in comparison with other benchmark allocation schemes.

11

Resource Provisioning to Spatio-Temporal Urban Services

Traditionally, resource allocation problems have been well studied in the Operations Research literature and are referred to as assignment problems. The classical bipartite assignment problem can be solved using the framework of Optimal Transportation [142], which is efficiently computed using Linear Programming for certain utility functions. Similarly, methods to solve the combinatorial optimization problems entailing to the assignment problem are available such as the Hungarian algorithm [74]. Modified versions of these methods involving convex optimization have been developed to solve resource allocation problems in several applications, e.g. radio resource management in wireless communications [53], taxi dispatch [94], data collection [8], etc. However, these techniques only present static and off-line solutions that cannot be implemented in real-time when there is incomplete information about future arrivals. Those that consider stochastic arrivals do not take the spatial aspect into account.

11.1 Spatio-Temporal Modeling of Urban Service Requests

Consider a typical source node at the origin with a discrete number of identical resources denoted by $N \in \mathbb{Z}^+$. The source is assumed to have an omni-directional service range of $R \in \mathbb{R}^+$. Service requests dynamically appear inside the served region according to a spatio-temporal Poisson process with intensity $\lambda(r, \theta, t) : [0, R] \times [0, 2\pi] \times \mathbb{R}^+ \to \mathbb{R}^+$, where the pair (r, θ) represents the polar coordinates of the service requests. Each service request is characterized by the tuple (X_i, D_i), where $X_i \in \mathbb{R}^+$ denotes the magnitude of the request and $D_i \in [0, R]$ denotes its distance from the source node. It is assumed that both the distance and the magnitude of requests are known at the source. Further, we assume that once a resource is allocated to service request, it becomes unavailable for allocation in the future.

Resource Management for On-Demand Mission-Critical Internet of Things Applications, First Edition. Junaid Farooq and Quanyan Zhu. © 2021 John Wiley & Sons, Inc. Published 2021 by John Wiley & Sons, Inc.

11.1.1 Characterization of Service Requests

We assume a time slotted[1] system with $\mathbf{t} = [1, 2, \ldots, T-1, T]$, where each time slot of duration τ represents an allocation period, also referred to as the decision horizon. To avoid non-uniformity in the average number of requests within allocation periods, we assume that the requests are uniform in the temporal domain, i.e. $\lambda(r, \theta, t) = \tau \tilde{\lambda}(r, \theta)$, where $\tau \in \mathbb{R}^+$ is a constant. Note that this assumption is not restrictive, since the temporal domain can be sliced into multiple regions of relatively uniform intensity and the framework can be modified accordingly. In each time slot, the service requests are distributed spatially according to a Poisson point process (PPP). It implies that the number of requests in a circle of radius R during one allocation period, denoted by K follows a Poisson distribution with average density Λ. The probability mass function (PMF) of K can be expressed as follows:

$$\mathbb{P}(K = k) = e^{-\Lambda} \frac{(\Lambda)^k}{k!}, \quad k = 0, 1, 2 \ldots, \tag{11.1}$$

where the average density of service requests during each allocation period, Λ, can be evaluated as follows:

$$\Lambda = \mathbb{E}[K] = \int_0^{2\pi} \int_0^R \tau \tilde{\lambda}(r, \theta) r \, dr \, d\theta. \tag{11.2}$$

Each service request has a type that represents the severity, criticality, or magnitude of demand. We model it as independent and identically distributed (i.i.d.)[2] random variables $X_i, i \in \{1, \ldots, K\}$, for each allocation period. Furthermore, we assume that the probability density function (pdf), denoted by $f_X(x)$, and cumulative distribution function (cdf), denoted by $F_X(x)$, of the magnitude of request are known at the source.[3] In order to exclude trivial cases in the resource allocation problem, we exclude the possibility of having no service request during an allocation period. It is stated formally stated as follows. It is assumed that there is at least one service request in every allocation period. Therefore we use the zero-truncated Poisson distribution to characterize the PMF of the number of service requests as follows:

$$\mathbb{P}(K = k | K > 0) = \frac{e^{-\Lambda}(\Lambda)^k}{(1 - e^{-\Lambda})k!}, \quad k = 1, 2, \ldots. \tag{11.3}$$

From the perspective of the source node, the distance of each service requests during an allocation period is also a random variable, which is independent of

---

1 Note that the discretization has been done to facilitate decision-making. However, the choice of allocation period is arbitrary and only impacts the granularity of the solution.

2 We assume independence across the temporal domain for the sake of simplicity of analysis. However, in practice, the magnitude of requests may exhibit temporal dependence, which has to be incorporated into the analysis.

3 In practical situations, the statistical information about service requests can be obtained using spatio-temporal estimation techniques [89].

the magnitude. The probability distribution of the distances can be expressed by the following lemma.

Lemma 11.1 *The pdf of the distance D, of a randomly selected service request inside a circular region of radius R from the source node, can be expressed as follows:*

$$f_D(d) = \frac{\int_0^{2\pi} d\tilde{\lambda}(d,\theta)d\theta}{\int_0^{2\pi} \int_0^R \tilde{\lambda}(r,\theta)r\, dr\, d\theta}. \tag{11.4}$$

Proof: The cdf of the distance to a typical service request in a non-homogeneous PPP model with density $\tau\tilde{\lambda}(r,\theta)$ can be evaluated as follows:

$$F_D(d) = \mathbb{P}(D \le d) = \frac{\int_0^{2\pi} \int_0^d \tilde{\lambda}(r,\theta)r\, dr\, d\theta}{\int_0^{2\pi} \int_0^R \tilde{\lambda}(r,\theta)r\, dr\, d\theta}.$$

The pdf can then be obtained as follows:

$$f_D(d) = \frac{dF_D(d)}{dd} = \frac{\int_0^{2\pi} d\tilde{\lambda}(d,\theta)d\theta}{\int_0^{2\pi} \int_0^R \tilde{\lambda}(r,\theta)r\, dr\, d\theta}.$$

In the special case of a homogeneous PPP, i.e. $\tilde{\lambda}(r,\theta) = \tau\lambda$, the cdf and pdf of the distance can be expressed as follows:

$$F_D(d) = \mathbb{P}(D \le d) = \frac{\tau\lambda\pi d^2}{\tau\lambda\pi R^2} = \frac{d^2}{R^2}, \tag{11.5}$$

$$f_D(d) = \frac{dF_D(d)}{dd} = \frac{2d}{R^2}. \tag{11.6}$$

$\square$

11.1.2 Utility of Resource Allocation

We assume a generic utility function that characterizes the benefit obtained by allocating a resource to a service request of magnitude X_i that is located at a distance of D_i from the source. The utility is denoted by $U(X_i, D_i) : \mathbb{R}^+ \times [0, R] \to \mathbb{R}^+$. The utility function is assumed to be monotonically increasing in X_i and monotonically decreasing in D_i,[4] since allocating a resource at a higher distance may incur additional cost. If the utility of obtained by allocating a resource to each service request during a single time slot is denoted by $Z_i = U(X_i, D_i)$, then Z_i is also a random variable with the i.i.d. property formally expressed in the following remark.

Remark 11.1 The random variables $Z_i = U(X_i, D_i)$, $i = 1, \ldots, K$, are i.i.d. random variables since X_i and D_i are i.i.d., respectively.

---

4 It implies that the utility is higher for close and high magnitude requests and vice versa.

For notational convenience, we will henceforth drop the index i and refer to the random variables describing the utility as Z. The cdf and pdf of the random variable Z can be evaluated as follows:

Lemma 11.2 *The cdf and pdf of the random variable Z describing the utility of allocation to a randomly selected service request can be evaluated, respectively, as follows:*

$$F_Z(z) = \mathbb{P}[Z \leq z] = \int\int_S f_X(x)f_D(d)dx\,dd \tag{11.7}$$

and

$$f_Z(z) = \frac{d}{dz}\int\int_S f_X(x)f_D(d)dx\,dd, \tag{11.8}$$

where $S = \{(x,d) : U(x,d) \leq z\}$.

We define the maximum utility during each time slot $j \in \{1, \dots, T\}$ as follows:

$$\tilde{Z}_j = \max_{1 \leq i \leq K_j} Z_i, \tag{11.9}$$

where $K_j, j = 1, \dots, T$ are i.i.d. Poisson random variables with mean Λ. Since the random variables $\tilde{Z}_j, j \in \{1, \dots, T\}$, for all time slots are also i.i.d., we hereby drop the subscript and refer to it as $\tilde{Z}$. Then using extreme value theory, the pdf of $\tilde{Z}$ can be expressed by the following lemma:

Lemma 11.3 *The pdf of $\tilde{Z} = \max\{Z_1, Z_2, \dots, Z_K\}$, where $\{Z_i\}_{1 \leq i \leq K}$ are i.i.d. random variables with cdf $F_Z(z)$ and pdf $f_Z(z)$, and K is a Poisson random variable with mean Λ, can be expressed as follows:*

$$f_{\tilde{Z}}(z) = \frac{\Lambda f_Z(z)e^{\Lambda(F_Z(z)-1)}}{1 - e^{-\Lambda}}. \tag{11.10}$$

Proof: The distribution function of $\tilde{Z} = \max\{Z_1, Z_2, \dots, Z_K\}$ with K being a Poisson random variable can be obtained as follows:

$$
\begin{aligned}
F_{\tilde{Z}}(z|K = k) &= \mathbb{P}(\tilde{Z} \leq z|K = k), \\
&= \mathbb{P}(\max\{Z_1, Z_2, \dots, Z_K\} \leq z|K = k), \\
&= \mathbb{P}(Z_1 \leq z, Z_2 \leq z, \dots, Z_K \leq z|K = k), \\
&= \prod_{i=1}^{K} \mathbb{P}(Z_i \leq z) = \prod_{i=1}^{K} F_{\tilde{Z}}(z) = (F_Z(z))^K.
\end{aligned} \tag{11.11}
$$

Consequently, the conditional pdf of $\tilde{Z}$ can be expressed as follows:

$$f_{\tilde{Z}}(z|K=k) = \frac{dF_{\tilde{Z}}(z|K=k)}{dz} = k(F_Z(z))^{k-1}f_Z(z). \tag{11.12}$$

Finally, the pdf can be obtained as follows:

$$\begin{aligned} f_{\tilde{Z}}(z) &= \sum_{k=1}^{\infty} f_{\tilde{Z}}(z|K=k)\mathbb{P}(K=k|K>0), \\ &= \sum_{k=1}^{\infty} k(F_Z(z))^{k-1}f_Z(z)\frac{e^{-\Lambda}(\Lambda)^k}{(1-e^{-\Lambda})k!}, \\ &= \frac{\Lambda f_Z(z)e^{\Lambda(F_Z(z)-1)}}{1-e^{-\Lambda}}. \end{aligned} \tag{11.13}$$

11.1.3 Problem Definition

The source has a total of T allocation periods during which it needs to allocate all the $N \leq T$ available resources to incoming requests. Instead of delaying allocation decision until all the requests have appeared, the goal is to decide allocation in real-time. Therefore, a mechanism is required to allocate resources dynamically while maximizing the total expected utility obtained from allocation. During each allocation period, the decision problem is whether to allocate a resource to one of the current utility maximizing requests or to wait for the next batch of requests to decide.

The resource provisioning problem can be formally expressed as follows:

$$\max_{\{i_1,i_2,\ldots,i_T\}\in\xi} \mathbb{E}\left[\sum_{j=1}^{T} \tilde{Z}_j q_{i_j}\right] \tag{11.14}$$

where ξ is the set of all possible permutations of the integers $1,\ldots,T$ and the vector $q = [q_1, q_2, \ldots, q_T]$ is such that $q_1 = q_2 == q_N = 1$ and $q_{N+1} = q_{N+2} = \ldots = q_T = 0$. The objective of the optimization is to find the permutation vector $[i_1, i_2, \ldots, i_T]$ that maximizes the expected total utility. The challenge lies in the fact that at every time slot, the decision has to be made in real-time without knowing the actual realizations of future request magnitudes.

11.2 Optimal Dynamic Allocation Mechanism

In this section, we describe the process to solve the optimization problem expressed in (11.16). However, we would first like to emphasize the fact that the decision problem is only relevant if $N \leq T$, i.e. the number of available resources is less than the number of allocation periods. In other words, if the number of resources is more than the number of allocation periods, then the optimal policy would be to allocate a resource to the utility maximizing request in every allocation period. A grid representing the possible pairs of (T, N) is illustrated in Figure 11.2.

11.2.1 Dynamic Programming Solution

We denote the optimal[5] value obtained by solving the optimization problem in (11.16) if T allocation periods are remaining and N resources are available as $V(T, N)$. Let $\rho_T^N \in \mathbb{R}^+$ denote the decision threshold on the random variable $\tilde{Z}$ if T allocation periods are remaining and N resources are available. Then, the value function can be expressed recursively using the following lemma:

Lemma 11.4 *If T allocation periods are remaining and N resources are available, then the total value obtained can be recursively expressed as follows:*

$$V(T, N) = \dots \begin{cases} \tilde{Z}_T + \mathbb{E}[V(T-1, N-1)], & \text{if } \tilde{Z} \geq \rho_T^n, \\ \mathbb{E}[V(T-1, N)], & \text{if } \tilde{Z} < \rho_T^n. \end{cases} \quad (11.15)$$

Proof: To prove the recursive value function, we make use of the decision tree shown in Figure 11.1. At the current allocation period, if there are T allocation

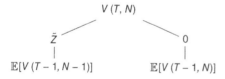

Figure 11.1 Decision tree when T allocation periods are remaining and N resources are available.

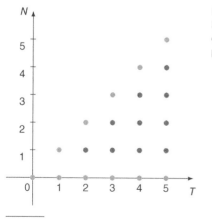

Figure 11.2 Light gray dots indicate the boundary cases for the pair (T, N). Dark gray dots indicate the cases for which the decision problem needs to be solved for allocation.

---

5 The term "optimal is used in an average sense, i.e. the maximum expected value is obtained in the long run if the proposed policy is implemented.

periods remaining with N available resources, there are two possible decisions, i.e. to allocate a resource to the utility maximizing request or to postpone allocation to a future allocation period. The value function can be expressed as follows:

$$V(T,N) = \max_{q_T}\{\tilde{Z}q_T + \mathbb{E}[V(T-1,N-1)], \ \mathbb{E}[V(T-1,N)]\}. \quad (11.16)$$

It is clear that if $q_T = 1$ and $\tilde{Z}_T$ is greater than some threshold ρ_T^N, then $\tilde{Z} + \mathbb{E}[V(T-1,N-1)] \geq \mathbb{E}[V(T-1,N)]$. Therefore, $V(T,N) = \tilde{Z} + \mathbb{E}[V(T-1,N-1)]$. Otherwise, if $q_T = 0$, then $\mathbb{E}[V(T-1,N)] > \mathbb{E}[V(T-1,N-1)]$, which results in $V(T,N) = \mathbb{E}[V(T-1,N)]$. $\square$

Using this value function, the optimal allocation thresholds can be obtained using the procedure provided in the following theorem.

Theorem 11.1 *If there are T allocation periods and N homogeneous resources available, then it is optimal to allocate an available resource to a utility maximizing request, characterized by $\tilde{Z}_T$, i.e. $q_T = 1$, if*

$$\tilde{Z}_T \geq \rho_T^n, \quad (11.17)$$

where the thresholds ρ_T^N can be obtained as

$$\rho_T^n = \mathbb{E}[V(T-1,n)] - \mathbb{E}[V(T-1,n-1)], \quad (11.18)$$

and the expected value functions with T allocation periods and N resources can be computed recursively as follows:

$$\mathbb{E}[V(T,n)] = \int_{\rho_T^n}^{\infty} (\tilde{Z} + \mathbb{E}[V(T-1,n-1)])f_{\tilde{Z}}(z)\, dz$$
$$+ \mathbb{E}[V(T-1,n)]F_{\tilde{Z}}(\rho_T^n). \quad (11.19)$$

Proof: From the decision tree shown in Figure 11.1, the decision to allocate a resource to a qualifying request is only made if the utility obtained from the allocation is higher than the utility obtained from postponing the decision. It implies that

$$\tilde{Z} + \mathbb{E}[V(T-1,n-1)] \geq \mathbb{E}[V(T-1,n)]. \quad (11.20)$$

This leads to the condition that

$$\tilde{Z} \geq \mathbb{E}[V(T-1,n)] - \mathbb{E}[V(T-1,n-1)] = \rho_T^N. \quad (11.21)$$

The expectation of the value function can be obtained directly from the definition in (11.4) in terms of the allocation threshold ρ_T^N. $\square$

To complete the optimal recursive solution, we need to evaluate the boundary conditions, i.e. the expected value functions for the (T,N) pairs highlighted by the light gray dots in Figure 11.2. The expected values for such cases can be expressed by the following lemma.

Lemma 11.5 *The expected value obtained for the boundary cases of (T, N) can be evaluated as follows.*

$$\mathbb{E}[V(T, 0)] = 0, \tag{11.22}$$

$$\mathbb{E}[V(n, n)] = n\mathbb{E}[\tilde{Z}]. \tag{11.23}$$

Proof: It follows from Theorem 1 that $\rho_n^n = \mathbb{E}[V(n-1, n)] - \mathbb{E}[V(n-1, n-1)] = 0, \forall n \in \mathbb{Z}^+$. It means that if the number of allocation periods equals the number of available resources, then the resource should be allocated to the utility maximizing request during that period regardless of the utility. Using this fact and the definition of $V(T, N)$, the result for $\mathbb{E}[V(n, n)]$ can be proved inductively. □

11.2.2 Computation and Implementation

In this section, we explain the procedure to compute the optimal allocation thresholds and provide an overview on implementing the proposed framework. In the case where there is one resource and two remaining time slots, the value function can be written as follows:

$$V(T = 2, N = 1) = \begin{cases} \tilde{Z} & \text{if } \tilde{Z} \geq \rho_2^1, \\ \rho_2^1 & \text{if } \tilde{Z} < \rho_2^1, \end{cases} \tag{11.24}$$

where the threshold $\rho_2^1 = \mathbb{E}[V(1, 1)] = \mathbb{E}[\tilde{Z}]$. In the case that there is one resource and $T = 3$ time slots remaining, then the value function can be expressed as follows:

$$V(T = 3, N = 1) = \begin{cases} \tilde{Z}, & \text{if } \tilde{Z} \geq \rho_3^1, \\ \mathbb{E}[V(2, 1)], & \text{if } \tilde{Z} < \rho_3^1, \end{cases} \tag{11.25}$$

where $\rho_3^1 = \mathbb{E}[V(2, 1)] - \mathbb{E}[V(2, 0)] = \int_{\rho_2^1}^{\infty} z f_{\tilde{Z}}(z) dz + \rho_2^1 F_{\tilde{Z}}(\rho_2^1)$. However, ρ_2^1 is available from the previous step. Similarly, by computing $\mathbb{E}[V(3, 1)]$ enables the computation of ρ_3^2 and $V(3, 2)$. This process needs to be executed recursively to obtain the value functions and the thresholds for arbitrary number of allocation periods and available resources. The step-wise procedure for computing the allocation thresholds is summarized in Algorithm 11.1 and a summary of the implementation procedure is provided in Algorithm 11.2. Once the form of $U(X, D)$ is known, the optimal thresholds ρ_T^N translate to concentric surfaces, as shown in Figure 11.3, that can be used to compare the maximal utilities in each time slot based on the number of available resources and time slots remaining to decide on allocation.

Algorithm 11.1 Optimal Threshold Computation

1: **function** COMPUTATION OF THRESHOLDS
Require: $F_{\breve{Z}}(z), f_{\breve{Z}}(z), \mathbb{E}[\breve{Z}]$.
Require: $EV \leftarrow [0_{T \times N}], \rho \leftarrow [0_{T \times N}]$.
2: **for** $i = 1$ to T **do**
3:   **for** $j = 1$ to N **do**
4:    **if** $i = j$ **then**
5:     $EV(i,j) \leftarrow j \times \mathbb{E}[\breve{Z}]$.
6:    **end if**
7:   **end for**
8: **end for**
9: **for** $i = 1$ to T **do**
10:   **for** $j = 1$ to N **do**
11:    **if** $j = 1$ **then**
12:     $\rho(i,j) \leftarrow EV(i-1,j)$.
13:     $EV(i,j) = \int_{\rho(i,j)}^{\infty} \breve{Z} f_{\breve{Z}}(z)\, dz +$
  $EV(i-1,j) F_{\breve{Z}}(\rho(i,j))$.
14:    **else**
15:     $\rho(i,j) \leftarrow EV(i-1,j) - EV(i-1,j-1)$.
16:     Compute $EV(i,j)$ using (11.19).
17:    **end if**
18:   **end for**
19: **end for**
20: **end function**

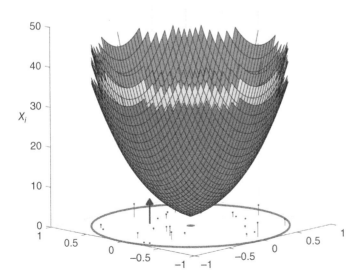

Figure 11.3 Spatial requests in one allocation period. The bold arrow represents the request with maximum magnitude. Concentric surfaces correspond to the allocation thresholds if one, two, and three resources are available, respectively.

11.3 Numerical Results and Discussion

In this section, we first present some special cases for which the optimal solution can be obtained numerically. Then we provide simulation results to demonstrate the practical implementation and effectiveness of the proposed framework. For sake of simple presentation of results, we assume a homogeneous spatio-temporal intensity of service requests, i.e. $\lambda(r, \theta, t) = \tau\lambda$. Then, service requests are distributed according to a Poisson process with intensity $\Lambda = \int_0^{2\pi}\int_0^R \lambda\tau r \, dr \, d\theta = \tau\lambda\pi R^2$. For analytical tractability and practical relevance, we will use two specific form of utility functions, i.e. $U(X, D) = X(1 + D)^{-\eta}$, $\eta \geq 0$, referred to as the *power law utility*[6], and $U(X, D) = Xe^{-\alpha D}$, referred to as the *exponential utility*.

11.3.1 Special Cases

11.3.1.1 Power Law Utility
In this section, we assume that the utility function is of the form of $U(X, D) = X(1 + D)^{-\eta}$, $\eta \geq 0$. We will further break down into two special cases, i.e. when the intensities of requests are exponentially and uniformly distributed.

Case I: Exponential Magnitude If $X \sim \exp(\mu^{-1})$, then $F_X(x) = 1 - e^{-\mu x}$, $x \geq 0$, and $f_X(x) = \mu e^{-\mu x}$, $x \geq 0$, then the pdf and cdf of the utility of each service request can be expressed by the following corollary.

Corollary 11.1 *The pdf and cdf of the utility Z of a service request inside a circular region of radius R if $U(X, D) = X(1 + D)^{-\eta}$ and $X \sim \exp(\mu^{-1})$, can be evaluated as follows:*

$$F_Z(z) = 1 - \frac{2}{\eta R^2}(E_{\frac{\eta-2}{\eta}}(z\mu) - (1 + R)^2 E_{\frac{\eta-2}{\eta}}(z\mu(1 + R)^\eta)$$
$$- E_{\frac{\eta-1}{\eta}}(z\mu) + (1 + R)E_{\frac{\eta-1}{\eta}}(z\mu(1 + R)^\eta)), \tag{11.26}$$

where $E_n(x) = \int_1^\infty \frac{e^{-xt}}{t^n} dt$ is the generalized exponential integral.

$$f_Z(z) = \int_0^R \mu d(1 + d)^\eta e^{-\mu z(1+d)^\eta} dd. \tag{11.27}$$

---

6 Power law models are commonly used to model the propagation of wireless signals.

Algorithm 11.2 Spatio-Temporal Resource Allocation

1: **procedure** RUNTIME
Require: T, N, ρ.
2:    **while** $T \geq 0$ **do**
3:        Obtain the tuple $(X_i, D_i)_{1 \leq i \leq K}$.
4:        Compute the utility $U(X_i, D_i)$ for all requests.
5:        Determine the maximal utility $\tilde{Z}$ using (11.9).
6:        **if** $\tilde{Z} \geq \rho(T, N)$ **then**
7:            Allocate a resource to request corresponding to $\tilde{Z}$.
8:            $N \leftarrow N - 1$
9:        **else**
10:            Skip allocation in the current allocation period.
11:        **end if**
12:        $T \leftarrow T - 1$
13:    **end while**
14: **end procedure**

Proof: The pdf of Z can be evaluated as follows:

$$F_Z(z) = \mathbb{P}[Z \leq z] = \mathbb{P}[X(1+D)^{-\eta} \leq z],$$
$$= \mathbb{P}[X \leq z(1+D)^{\eta}]. \tag{11.28}$$

This probability can be obtained by integrating the joint density $f_{X,D}(x, d)$ over the shaded region $S = \{(x, d) : U(x, d) \leq z\}$ shown in Figure 11.4a. Consequently, the cdf can be expressed as follows:

$$F_Z(z) = \int_0^R \int_0^{z(1+d)^{\eta}} f_X(x) f_D(d) dx\, dd,$$
$$= \int_0^R \int_0^{z(1+d)^{\eta}} \mu e^{-\mu x} \times 2\frac{d}{R^2} dx\, dd. \tag{11.29}$$

Computing the integrals results in the expression provided in Corollary 11.1. Furthermore, differentiating $F_Z(z)$ w.r.t. z results in $f_Z(z)$. $\qquad \square$

Case II: Uniform Magnitude If $X \sim \text{Unif}(0, \beta)$, then $F_X(x) = \frac{x}{\beta}$, $x \in (0, \beta)$ and $f_X(x) = \frac{1}{\beta}$, $x \in (0, \beta)$, then the pdf and cdf of the utility of each service request can be expressed by the following corollary.

Corollary 11.2 *The pdf and cdf of the utility Z of a service request inside a circular region of radius R if $U(X,D) = X(1+D)^{-\eta}$ and $X \sim U(0, \beta)$, can be evaluated as follows:*

$$F_Z(z) = \begin{cases} \dfrac{2z}{\beta R^2}\left(\dfrac{1+\left((\frac{\beta}{z})^{1/\eta}\right)^{\eta+1}(R(\eta+1)-1)}{2+3\eta+\eta^2}\right), & \text{if } 0 \leq z \leq \beta(1+R)^{-\eta}, \\[4ex] \dfrac{2z}{\beta R^2}\left(\dfrac{1+(R+1)^{\eta+1}\left((\eta+1)\left(\left(\frac{\beta}{z}\right)^{(1/\eta)}\right)-\eta 2\right)}{2+3\eta+\eta^2}\right) \\[2ex] \quad +\dfrac{2}{R^2}\left(\dfrac{R^2}{2}-\dfrac{1}{2}\left(\left(\dfrac{\beta}{z}\right)^{1/\eta}-1\right)^2\right), & \text{if } \beta(1+R)^{-\eta} < z \leq \beta, \end{cases}$$

and

$$f_Z(z) = \begin{cases} \dfrac{2}{\beta R^2}\left(\dfrac{1+(R+1)^{\eta+1}(R(\eta+1)-1)}{2+3\eta+\eta^2}\right), & \text{if } 0 \leq z \leq \beta(1+R)^{-\eta}, \\[4ex] \dfrac{1}{\beta R^2 \eta}\left(2\eta\left(1-2(\frac{\beta}{z})^{1+2/\eta}+\left(\dfrac{\beta}{z}\right)^{\frac{\eta+1}{\eta}}\right)\right. \\[2ex] \left.-4\left(\dfrac{\beta}{z}\right)^{\frac{\eta+1}{\eta}}\left(\left(\dfrac{\beta}{z}\right)^{1/\eta}-1\right)\right), & \text{if } \beta(1+R)^{-\eta} < z \leq \beta. \end{cases}$$

$$(11.30)$$

Proof: The pdf of Z can be evaluated as follows:

$$F_Z(z) = \mathbb{P}[Z \leq z] = \mathbb{P}[X(1+D)^{-\eta} \leq z],$$
$$= \mathbb{P}[X \leq z(1+D)^{\eta}]. \qquad (11.31)$$

This probability can be obtained by integrating the joint density $f_{X,D}(x,d)$ over the shaded region $S = \{(x,d) : U(x,d) \leq z\}$ shown in Figure 11.4b.

$$F_Z(z) = \begin{cases} \int_0^R \int_0^{z(1+d)^\eta} \frac{1}{\beta} \times 2\frac{d}{R^2}dx\, dd, & \text{if } z \leq \beta(1+R)^{-\eta}, \\[3ex] \int_0^{\left(\frac{\beta}{z}\right)^{\frac{1}{\eta}}-1} \int_0^{z(1+d)^\eta} \frac{1}{\beta} \times 2\frac{d}{R^2}dx\, dd \\[2ex] \quad +\int_{\left(\frac{\beta}{z}\right)^{\frac{1}{\eta}}-1}^{R} \int_0^{\beta} \frac{1}{\beta} \times 2\frac{d}{R^2}dx\, dd, & \text{if } \beta(1+R)^{-\eta} < z \leq \beta. \end{cases}$$

$$(11.32)$$

Computing the integrals leads to the result in Lemma 11.2. The density function of Z can then be evaluated by differentiating $F_Z(z)$ w.r.t. z. $\qquad \square$

11.3.1.2 Exponential Utility
In this section, we assume that the utility function is of the form of $U(X,D) = Xe^{-\alpha D}$, $\alpha \geq 0$. We will further break down into two special cases, i.e. when the intensities of requests are exponentially and uniformly distributed.

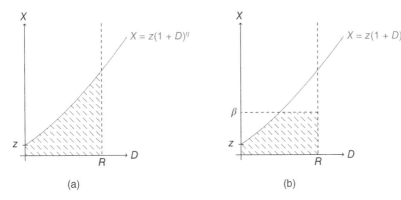

Figure 11.4 Resource allocation thresholds for exponential magnitude of service requests. (a) Power law utility with exponential magnitude. (b) Exponential utility with exponential magnitude.

Case I: Exponential Magnitude If the magnitude of requests follows an exponential distribution with mean μ^{-1}, then the pdf and cdf of the utility of each service request can be expressed by the following corollary.

Corollary 11.3 *The pdf and cdf of the utility Z of a service request inside a circular region of radius R if $Xe^{-\alpha D}$, $\alpha \geq 0$ and $X \sim \exp(\mu^{-1})$, can be evaluated as follows:*

$$F_Z(z) = 1 - \frac{2}{R^2} \int_0^R de^{-\mu z e^{\alpha d}} \, dd \tag{11.33}$$

and

$$f_Z(z) = \int_0^R \frac{2\mu d}{R^2} e^{\alpha d - \mu z e^{\alpha d}} \, dd. \tag{11.34}$$

Proof: The proof follows a similar methodology as used in the proof of Corollary 11.1 and has been omitted for brevity. □

Case II: Uniform Magnitude If the magnitude of requests follows a uniform distribution in $(0, \beta)$, then the pdf and cdf of the utility of each service request can be expressed by the following corollary.

Corollary 11.4 *The pdf and cdf of the utility Z of a service request inside a circular region of radius R if $Xe^{-\alpha D}$, $\alpha \geq 0$ and $X \sim U(0, \beta)$, can be evaluated as follows:*

$$F_Z(z) = \begin{cases} \dfrac{2z}{\alpha^2 \beta R^2}(1 + e^{\alpha R}(\alpha R - 1)), & if > \ 0 \leq z \leq \beta e^{-\alpha R}, \\[2ex] \dfrac{2}{\alpha^2 \beta R^2}(z + z e^{\alpha R}(\alpha R - 1)), & if \ \beta e^{-\alpha R} < z \leq \beta, \end{cases} \tag{11.35}$$

$$f_Z(z) = \begin{cases} \dfrac{2}{\alpha^2 \beta R^2}(1 + e^{\alpha R}(\alpha R - 1)), & \text{if } 0 \le z \le \beta e^{-\alpha R}, \\[12pt] \dfrac{2}{\alpha^2 \beta R^2}(1 + e^{\alpha R}(\alpha R - 1)), & \text{if } \beta e^{-\alpha R} < z \le \beta. \end{cases} \quad (11.36)$$

Proof: The proof follows a similar methodology as used in the proof of Corollary 11.2 and has been omitted for brevity. □

11.3.2 Performance Evaluation and Comparison

To illustrate the performance of our developed resource provisioning framework, we conduct simulation experiments. The model parameters, unless otherwise stated, are selected as follows for illustrative purposes: service range $R = 1$, service request density $\lambda = 10$ requests per unit area, duration of each time slot $\tau = 1$, number of time slots $T = 30$, number of available resources $N = 10$, mean of exponential magnitude $\mu^{-1} = 1$, and parameter of uniform magnitude $\beta = 1$. Service requests are generated according to the homogeneous spatio-temporal Poisson process in a circular region around the origin. At each successive time slot, the maximum utility is picked and a resource is allocated to the corresponding request only if the utility is higher than the threshold. Otherwise the requests are discarded and the next time slot is observed until the terminal time is reached.

Figure 11.5 plots the optimal allocation thresholds against the number of allocation periods remaining for different number of available resources at the source node if the magnitude is assumed to be exponentially distributed. The threshold successively increases as more allocation periods are available for the same number of available resources. In other words, the source becomes more selective in allocation as more allocation periods are available. Similarly, for a given number of allocation periods, the threshold decreases as the number of available resources increases. Note that the threshold remains zero if the number of available resources and remaining allocation periods are equal. The successive difference between the thresholds is attributed to the fact that the magnitude of requests is exponentially distributed. A similar behavior will be observed in the case of uniformly distributed arrivals except that the difference between successive thresholds is expected to be uniform.

We also compare our proposed resource provisioning framework with benchmark schemes, namely, the Ideal allocation, myopic allocation, and random allocation. The Ideal allocation represents the case when there is no uncertainty about the future and the maximum utilities in each time slot are known *a priori*. Therefore, it results in the best possible resource allocation strategy. The Myopic allocation strategy allocates a resource to the utility maximizing request in each time slot regardless of its magnitude. The random allocation is a generalization of the myopic allocation where in every time slot,

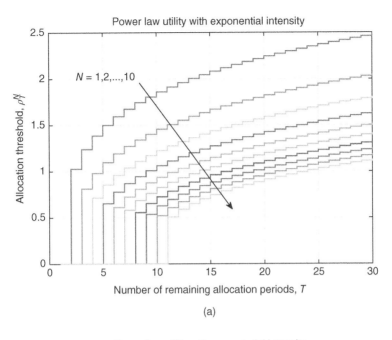

(a)

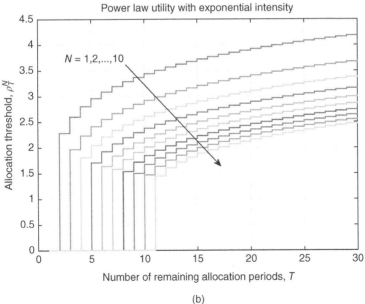

(b)

Figure 11.5 Total expected utility against varying spatio-temporal density of requests.

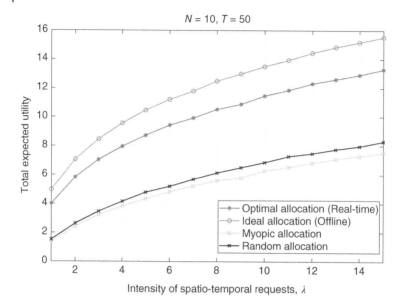

Figure 11.6 Total expected utility against varying magnitude of requests.

a resource is allocated to the utility maximizing request with a probability of 0.5. If the probability of allocation is further reduced, there may be an increase in total expected utility since there is a higher chance of hitting better utility requests. Figures 11.6 and 11.7 plot the total expected utility of allocation under varying magnitude of spatio-temporal requests and the mean magnitude of requests, respectively. The simulation results clearly show that the proposed optimal allocation strategy lower expected utility as compared with the Ideal allocation. The loss in utility is due to the fact that allocation decisions are made in real-time without information about future arrival of requests. However, it performs at least twofold better than the myopic and random allocation strategies (Figure 11.7).

11.4 Summary and Conclusions

In this chapter, we have proposed a utility maximizing approach to allocate resources from a centralized source location to spatio-temporal service requests with varying magnitude of demand. The framework is highly generic in terms of the utility of allocation, the number of available resources, and the number of allocation periods along with the spatio-temporal statistics of the requests. Using statistical analysis of the utility obtained by the allocation, we have developed an optimal filtering scheme that makes only qualifying

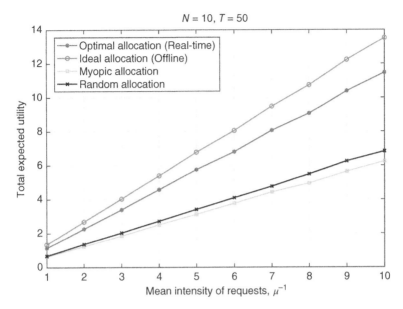

Figure 11.7 The shaded region represents the region of integration $S = \{(x, d) : U(x, d) \leq z\}$. (a) Shaded region for power law utility and exponential magnitude. (b) Shaded region for power law utility and uniform magnitude.

requests eligible for resource allocation. The developed resource provisioning framework is envisioned to have wide ranging applications in smart cities.

Several useful extensions can be done to enhance the proposed framework to cater for more realistic and tailored scenarios for different resource allocation problems. The cost of delaying the allocation can be added as a parameter to the framework to prevent over-selectiveness of the source. Furthermore, the case of non-homogeneous resources and the continuous time version of the framework can be investigated to cater for more versatile scenarios and applications.

Part VI

Conclusion

12

Challenges and Opportunities in the IoT Space

12.1   Broader Insights and Future Directions

This book takes a clean-slate approach toward designing intelligent decision mechanisms that enable autonomous mission-critical service delivery. In the section, some of the broader insights that can be obtained from the results of this book are presented.

12.1.1   Distributed Cross-Layer Intelligence for Mission-Critical IoT Services

Autonomous operation of Cyber-Physical Systems (CPS)/Internet of things (IoT) systems requires an interdisciplinary and cross-layer approach due to the coupling between cyber and physical components (Figure 12.1). The state and communication at the cyber layer influences the dynamics and control at the physical layer and vice versa. Hence, a cross-layer and distributed approach is crucial for effective design and operation of such systems, particularly for massive systems-of-systems scenarios. The following subsections summarize the key insights of this work.

12.1.1.1   Secure and Resilient Networking for Massive IoT Networks

The fundamental goal of the IoT is to inter-connect smart objects so that they can exchange data and leverage the capabilities of each other to achieve individual and/or network objectives such as high situational awareness, efficiency, and accuracy. This connectivity relies on wireless communication networks, which have limitations based on the communication technologies involved. However, oftentimes the connectivity is challenged by limitations such as lack of communication infrastructure or limited available wireless spectrum. In Part III, of the book, we have tackled the physical connectivity challenge using

Resource Management for On-Demand Mission-Critical Internet of Things Applications, First Edition.
Junaid Farooq and Quanyan Zhu.
© 2021 John Wiley & Sons, Inc. Published 2021 by John Wiley & Sons, Inc.

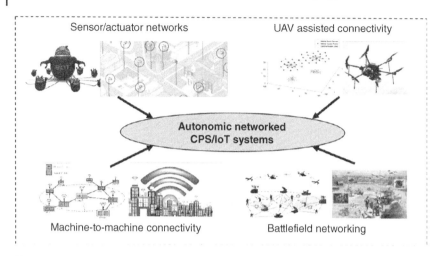

Sensor/actuator networks — UAV assisted connectivity

Autonomic networked CPS/IoT systems

Machine-to-machine connectivity — Battlefield networking

Figure 12.1 Autonomic CPS/IoT systems.

a multitude of approaches, such as device-to-device (D2D) communication in vehicular networks, augmentation of wireless connectivity using a network of unmanned aerial vehicles (UAVs), and ensuring spectrum availability via reservation contracts. The work on augmented wireless connectivity with the assistance of UAVs uses a feedback system approach to dynamically configure the network ensuring resilient connectivity in spatially dispersed wireless devices under the possibility of cyber–physical attacks. It is one of the first works in literature to use dynamic configurability as opposed to static placement optimization. Finally, the research on spectrum reservation uses an economic approach to provide guaranteed spectrum access for mission-critical Internet of things (MC-IoT) applications.

12.1.1.2 Autonomic Networked CPS: From Military to Civilian Applications

Connected CPS networks are used for dissemination of data and information for enhanced situational awareness and decision-making. Hence, it is imperative to design the networks in a way that ensures that information reliably propagates throughout the network and also ensures that networks have the capability to recover from failures/attacks that may sabotage operation. Of particular interest are the military and tactical networks with stringent requirements such as support for extreme heterogeneity, rapid reconfigurability, and mosaic warfare needs. In this respect, one of the key contributions of this book has been the development of a secure and reconfigurable network design framework suitable for adversarial environments such as the Internet of battlefield things (IoBT). The readers are directed to Chapter 7 for details. Networks with enhanced data dissemination capabilities also open doors for malicious activity and malware. Furthermore, malicious entities such as compromised supply chain actors may exploit backdoor channels for stealthy takeover and cause large scale coordinated attacks. To tackle this security risk posed by wireless connectivity proliferation in IoT, this book has developed a mechanism for inspecting devices based on their wireless neighborhood that prevents a large scale coordinated attack while causing minimum operational interruption. This work is one of the first works on dealing with botnets using the wireless interface as opposed to the traditional Internet based infiltration. This work also brings together theories from spatial point processes, mathematical epidemiology, control theory, and optimization to advance the understanding of complex malware processes over wireless networks and their effective control.

12.1.1.3 Strategic Resource Provisioning for Mission-Critical IoT Services

Limited available resources need to be effectively allocated in an IoT ecosystem at various different levels, particularly for delivering MC-IoT services. At the communication layer, the spectrum resources need to be effectively provisioned to applications according to their performance requirements and power limitations. Similarly, at the cloud layer, computing nodes and data processing resources need to be allocated and priced strategically to ensure maximum revenue for the cloud service provider amidst uncertainty of service requests. In this regard, in Part V, this book has developed a real-time resource allocation and pricing framework for cloud-enabled IoT systems, where the computational complexity of arriving tasks is evaluated and is accordingly assigned to the available computing bundles. Similarly, for low latency applications, a framework is developed to appropriately select the resources available at one of the fog/edge computing nodes for real-time tasks. These works extend the existing works on cloud computing and 5G ultra low latency applications by incorporating the real-time aspect. Furthermore, the allocation and pricing framework is extended to the case where there is uncertainty in the spatial domain in addition to the temporal domain. The policies can also be learned from available data using techniques such as reinforcement learning. This is useful in decision-making for various IoT-enabled smart city applications such as in addressing real-time emergency response and public safety requests.

12.2 Future Research Directions

This section describes some of the potential research directions and extensions that are promising in pushing the boundaries of this research further to encompass more impactful real world applications. A broader vision and a brief description of the methodologies involved are presented to act as a guideline for future research (Figure 12.2).

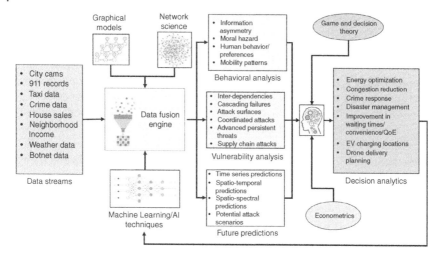

Figure 12.2 Overview of future research and approach.

12.2.1 Distributed Learning and Data Fusion for Security and Resilience in IoT-Driven Urban Applications

The availability of contextually rich data streams in the urban setting along with advances in machine learning and computational capabilities provides an untapped opportunity to analyze and learn human behaviors. It allows us to understand, preferences, incentives, and relationships that act as a basis to develop mechanisms for achieving desired goals and objectives. Future work can focus on leveraging urban data along with theoretical models to develop data-driven mechanisms that will pave the way for the realization of safe and resilient smart cities. Some of the possible research directions are detailed in the following subsections.

12.2.1.1 Data-Driven Learning and Decision-Making for Smart City Service Provisioning

IoT-driven urban services are rapidly emerging in almost all industry verticals. Companies like Uber, lyft, Via, etc. have come up with a range of on-demand urban mobility solutions and are moving toward autonomous microtransit solutions. Similarly, there is a strong interest in autonomous vehicle and drone based delivery services. These applications require intelligent decision-making by autonomous agents. Future research can identify ways to unlock the potential of publicly available urban data sets to improve the efficiency, security, and resilience decisions in smart urban infrastructure systems. The research questions that can be investigated are the following: (i) Can we use the massive urban data generated to assist in making operational decisions? (ii) Can we use data to explain hidden patterns and behaviors and paradoxes such as the tragedy of commons, moral hazard (exploitation due to information asymmetry), human preferences, disease causation, etc.? What are the safety and security threats? Several data-sets are available such as urban taxi data, city webcams, house sales, crime data, weather data, etc. to study these research questions. Objectives such as safety and mobility are critical concerns as the metropolises are developing. The decision questions in this thrust are the following: (i) Can we make urban transportation safe and secure? What policy or decisions should be made to mitigate threats? (ii) Can we make urban systems resilient-by-design? (iii) Can we discover ways to reduce accidents from patterns in historical accidents?

12.2.1.2 Market Design for On-Demand and Managed IoT-Enabled Urban Services

The availability of interconnected CPS in an IoT ecosystem allows for on-demand and managed service models to emerge. Some example urban services are depicted in Figure 12.3. New service paradigms such as Sensing-as-a-Service, Actuation-as-a-Service, Security as a Service, Resilience as a Service, etc. are gaining interest of the industry. By 2020, more than 24 billion devices will be connected to the Internet and more than $6 trillion will be invested in IoT solutions and services. For such novel architectures, it is important to develop non-traditional mechanisms for revenue maximizing allocations and pricing, contract development, risk-transfer policies, etc. In the future, new service models can be developed based on models in the literature on contract theory, prospect theory, etc., tailored for the IoT-driven smart city ecosystem to demonstrate the potential and business case for technology driven solutions. Pricing and market design are key aspects that can be

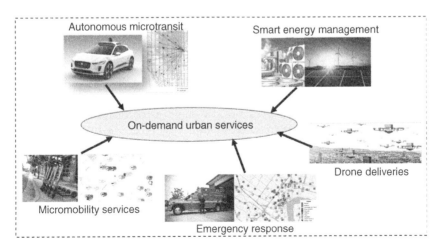

Figure 12.3 Examples of urban on-demand services.

investigated in this thrust. It requires econometric modeling of interactions in the data market along with human involvement through technologies such as crowd-sensing.

12.2.1.3 Proactive Resiliency Planning and Learning for Disaster Management in Cities

Emergency response and first responder tactical systems are key to the safety and resilience of smart cities. The use of sensing capabilities, historical data, and distributed data processing can not only assist early detection and rapid response, but can also help in emergency preparedness. For instance, wildfires are natural disasters that pose a significant threat to the metropolis in terms of life, health, safety, etc. The use of data from sensors reporting temperatures along with weather data can help in determining the timing and location of such incidents. Similarly, timely emergency response is critical for urban safety and quality of life. The average emergency response time across the United Stated is around 15 minutes, while in some states such as Wyoming, the average response time can be as high as 35 minutes. This may significantly elevate the losses and damages. My research will focus on the following questions: (i) Can we proactively select deployment locations of emergency responders or tactical units? (ii) Can safe routing and navigation be done based on crime data? (iii) Can cyber, physical, and social infrastructure be leveraged for disaster management and emergency response? The proposed approach is to use advanced analytics and data driven optimization methods that are distributionally robust and based on learning the generative model of the data to make decisions. Future research can focus on developing hybrid techniques based on model-driven decisions and learning components to improve the efficiency, safety, and resilience of urban intelligent infrastructure systems. An overview of the potential future research is illustrated in Figure 12.2.

12.2.2 Supply Chain Security and Resilience of IoT

Supply chain security is becoming an important factor in security risk analysis for modern information and communication technology (ICT) systems. As IoT devices proliferate and get adopted into critical infrastructure, the role of suppliers in risk assessment becomes all the more significant. IoT security risks are affected by supplier trust since suppliers possess the capacity to modify black box systems without detection. The risks posed by potentially malicious or compromised suppliers are compounded by interdependence among suppliers. Analyzing supply chain risks in IoT systems and networks is a crucial step in measuring the potential threat to the system and is a necessary precursor to making risk minimizing decisions regarding the supply chain. Supply chain risk management (SCRM) is a well-studied subject in literature [135]. However, supply chain risks in IoT systems are fundamentally different from traditional industries such as food and medicine [106]. Among other reasons, the opacity of black box systems renders assessment of risks considerably more difficult.

Furthermore, IoT supply chain risks are also different than the risks in traditional ICT systems due to their inter-connectivity. Suppliers themselves are inter-connected in ways that potentially increase risk, and these relationships are continually changing as firms adapt to market conditions. Hence, analyzing supply chain risks in IoT systems requires an adaptation of risk analysis techniques to consider these varied and complex sources of risk. The core challenge involved in supply chain threats is that suppliers may potentially alter the system's functions in indeterminate ways. Therefore, any component could be a potential vector for such a threat, not only those marked as important for security. Risk analysis can take this into account only by widening the class of components under question and including their suppliers. This broader approach requires supply chain risk analysis to leverage tools from system reliability theory. An example of a supply-chain attack would be when a supplier provides a product with a degraded implementation of a common cryptographic protocol, all the while claiming the product correctly uses the protocol to ensure the confidentiality of customer information. In the absence of extensive testing, the users of this product may assess the risk of using the product based on the (mis)information provided by the supplier. Not only does the product leave the customers open to attacks against the confidentiality of their data, the assessed risks are based on false information. Any mitigation efforts taken by the customer are significantly sub-optimal because of this supplier deception. Consequently, risk analysis must go beyond inherent security risks and take into account the trustworthiness of suppliers.

12.3   Concluding Remarks

This book is an attempt to build the theoretical underpinnings of decision science in IoT-enabled systems and networks, which are increasingly becoming important in all infrastructure. The developed models and theoretical approaches can act as a foundation to advance the science of data-driven decision-making in future urban infrastructure systems. Research in this direction will help in making our cities safer, sustainable, and more productive. The research will result in implementable algorithms and solutions that can be commercialized. It will have tremendous social and monetary impact attracting the interest of city administrations as well as private sector service providers in a smart city, e.g. micro transportation companies such as Uber, Lyft, Bird, Via, etc. Furthermore, enhanced preparedness and planning will improve disaster and emergency response leading to reduction in monetary and non-monetary losses. For instance, timely detection and response to wildfires can significantly reduce the loss of precious lives and huge monetary losses (around \$2.4 billion in 2017) in insurance payments. In addition to the measurable societal impact, this research will contribute toward training a workforce that invents the future of cities and is more empowered with the thinking and tools that will revolutionize and accelerate economic development. We believe that the work presented in this book will open up new dimensions for future researchers to build upon and enter into novel, exciting, and more impactful developments.

Bibliography

1 N. Abramson. THE ALOHA SYSTEM: another alternative for computer communications. In *Proceedings of the November 17–19, 1970, Fall Joint Computer Conference*, AFIPS '70 (Fall), pages 281–285, New York, NY, USA, 1970. ACM. http://doi.acm.org/10.1145/1478462.1478502.

2 H. Ahmadi, K. Katzis, and M. Z. Shakir. A novel airborne self-organising architecture for 5G+ networks. In *Proceedings of the IEEE Vehicular Technology Conference (VTC-Fall 2017)*, Toronto, Canada, Sept. 2017.

3 S. C. Albright. Optimal sequential assignments with random arrival times. *Management Science*, 21 (1): 60–67, 1974. ISSN 00251909, 15265501.

4 A. Al-Fuqaha, M. Guizani, M. Mohammadi, M. Aledhari, and M. Ayyash. Internet of things: a survey on enabling technologies, protocols, and applications. *IEEE Communication Surveys and Tutorials*, 17 (4): 2347–2376, Fourth Quarter 2015.

5 A. Al-Hourani, S. Kandeepan, and A. Jamalipour. Modeling air-to-ground path loss for low altitude platforms in urban environments. In *Proceedings of the IEEE Global Communications Conference (Globecom 2014)*, pages 2898–2904, Dec. 2014.

6 A. Al-Hourani, S. Kandeepan, and S. Lardner. Optimal LAP altitude for maximum coverage. *IEEE Wireless Communications Letters*, 3 (6): 569–572, Dec 2014.

7 S. Al-Sarawi, M. Anbar, K. Alieyan, and M. Alzubaidi. Internet of things (IoT) communication protocols: review. In *8th International Conference on Information and Technology (ICIT 2017)*, Amman, Jordan, May 2017.

8 I. Ali, A. Gani, I. Ahmedy, I. Yaqoob, S. Khan, and M. H. Anisi. Data collection in smart communities using sensor cloud: recent advances, taxonomy, and future research directions. *IEEE Communications Magazine*, 56 (7): 192–197, Jul. 2018.

9 M. Antonakakis, T. April, M. Bailey, M. Bernhard, E. Bursztein , J. Cochran, Z. Durumeric, J. A. Halderman, L. Invernizzi, M. Kallitsis, D. Kumar, C. Lever, Z. Ma, J. Mason, D. Menscher, C. Seaman, N. Sullivan, K. Thomas, Y. Zhou. Understanding the Mirai botnet. In *Proceedings of*

Resource Management for On-Demand Mission-Critical Internet of Things Applications, First Edition.
Junaid Farooq and Quanyan Zhu.
© 2021 John Wiley & Sons, Inc. Published 2021 by John Wiley & Sons, Inc.

the 26th USENIX Security Symposium, 2017. https://www.usenix.org/conference/usenixsecurity17/technical-sessions/presentation/antonakakis.

10 M. Armbrust, A. Fox, R. Griffith, A. D. Joseph, R. Katz, A. Konwinski, G. Lee, D. Patterson, A. Rabkin, I. Stoica, and M. Zaharia. A view of cloud computing. *ACM Communications*, 53 (4): 50–58, Apr. 2010. ISSN 0001-0782.

11 F. Baccelli and B. Baszczyszyn. Stochastic geometry and wireless networks: volume I theory. *Foundations and Trends in Networking*, 3 (3–4): 249–449, 2010. ISSN 1554-057X.

12 B. Banerjee, A. Mukherjee, M. K. Naskar, and C. Tellambura. BSMAC: a hybrid MAC protocol for IoT systems. In *IEEE Global Communications Conference (GLOBECOM 2016)*, Washington, DC, USA, Dec. 2016.

13 O. Bello and S. Zeadally. Intelligent device-to-device communication in the internet of things. *IEEE Systems Journal*, 10 (3): 1172–1182, Sept. 2016.

14 E. Bertino and N. Islam. Botnets and Internet of things security. *Computer*, 50 (2): 76–79, Feb. 2017. doi.ieeecomputersociety.org/10.1109/MC.2017.62.

15 F. Bonomi, R. Milito, J. Zhu, and S. Addepalli. Fog computing and its role in the internet of things. In *Proceedings of the 1st Edition of the MCC Wkshp Mobile Cloud Computing*, MCC '12, pages 13–16, New York, NY, USA, 2012. ACM. ISBN 978-1-4503-1519-7.

16 R. I. Bor-Yaliniz, A. El-Keyi, and H. Yanikomeroglu. Efficient 3-D placement of an aerial base station in next generation cellular networks. In *Proceedings of the IEEE International Conference on Communication (ICC 2016)*, pages 1–5, Kuala Lumpur, Malaysia, May 2016.

17 A. Brogi and S. Forti. QoS-aware deployment of IoT applications through the fog. *IEEE Internet of Things Journal*, 4 (5): 1185–1192, Oct. 2017.

18 R. L. Burden, J. D. Faires, and A. M. Burden. *Numerical Analysis*. PWS Publishers, 1985.

19 Can wireless LAN denial of service attacks be prevented? Understanding WLAN DoS vulnerabilities & practical countermeasures. Motorola Inc., White Paper, 2009.

20 H. Cai, B. Xu, L. Jiang, and A. V. Vasilakos. Iot-based big data storage systems in cloud computing: perspectives and challenges. *IEEE Internet of Things Journal*, 4 (1): 75–87, Feb. 2017.

21 X. Chen, L. Jiao, W. Li, and X. Fu. Efficient multi-user computation offloading for mobile-edge cloud computing. *IEEE/ACM Transactions on Networking*, 24 (5): 2795–2808, Oct. 2016.

22 L. Chen and H. Shen. Considering resource demand misalignments to reduce resource over-provisioning in cloud datacenters. In *Proceedings of the IEEE Conference on Computer Communications (INFOCOM 2017)*, May 2017.

23 J. Chen and Q. Zhu. Resilient and decentralized control of multi-level cooperative mobile networks to maintain connectivity under adversarial environment. In *IEEE 55th Conference on Decision and Control (CDC 2016)*, Las Vegas, USA, Dec. 2016.

24 M. Chiang and T. Zhang. Fog and IoT: an overview of research opportunities. *IEEE Internet of Things Journal*, 3 (6): 854–864, Dec. 2016.

25 F. Chiti, R. Fantacci, and B. Picano. A matching theory framework for tasks offloading in fog computing for IoT systems. *IEEE Internet of Things Journal*, 5 (6): 5089–5096, Dec. 2018.

26 P. Courty and L. Hao. Sequential screening. *The Review of Economic Studies*, 67 (4): 697–717, 2000.

27 Z. Cvetkovski. *Inequalities: Theorems, Techniques and Selected Problems*, chapter Convexity, Jensen's Inequality, pages 69–77. Springer-Verlag, Berlin, Heidelberg, 2012.

28 J. Dall and M. Christensen. Random geometric graphs. *Physical Review E: Covering Statistical, Nonlinear, Biological, and Soft Matter Physics*, 66: 016121, Jul. 2002. doi: 10.1103/PhysRevE.66.016121.

29 M. S. Daskin and K. L. Maass. *The p-Median Problem*, pages 21–45. Springer International Publishing, Cham, 2015.

30 F. C. Delicato, P. F. Pires, and T. Batista. *Resource Management for Internet of Things*, pages 45–104. Springer International Publishing, Cham, 2017. ISBN 978-3-319-54247-8. doi: 10.1007/978-3-319-54247-8_5.

31 R. Deng, R. Lu, C. Lai, T. H. Luan, and H. Liang. Optimal workload allocation in fog-cloud computing toward balanced delay and power consumption. *IEEE Internet of Things Journal*, 3 (6): 1171–1181, Dec. 2016.

32 C. Derman, G. J. Lieberman, and S. M. Ross. A sequential stochastic assignment problem. *Management Science*, 18 (7): 349–355, 1972. doi: 10.1287/mnsc.18.7.349.

33 M. De Sanctis, E. Cianca, G. Araniti, I. Bisio, and R. Prasad. Satellite communications supporting internet of remote things. *IEEE Internet of Things Journal*, 3 (1): 113–123, Feb. 2016.

34 Y. Dibrov. The Internet of things is going to change everything about cybersecurity. Harvard Business Review, Dec. 2017.

35 O. Dousse, M. Franceschetti, N. Macris, R. Meester, and P. Thiran. Percolation in the signal to interference ratio graph. *Journal of Applied Probability*, 43 (2): 552–562, 2006.

36 Cheng-yi Xia, Z. Wang, J. Sanz, S. Meloni, and Y. Moreno. Optimising police dispatch for incident response in real time. *Journal of the Operational Research Society*, 70 (2): 269–279, 2019.

37 S. Dunnett, J. Leigh, and L. Jackson. Effects of delayed recovery and nonuniform transmission on the spreading of diseases in complex networks. *Physica A: Statistical Mechanics and its Applications*, 392 (7): 1577–1585, 2013.

38 M. J. Farooq and Q. Zhu. Cognitive connectivity resilience in multi-layer remotely deployed mobile internet of things. In *IEEE Global Communications Conference (Globecom 2017)*, Singapore, Dec. 2017.

39 M. J. Farooq and Q. Zhu. Secure and reconfigurable network design for critical information dissemination in the internet of battlefield things (IoBT). In *15th International Symposium on Modeling and Optimization in Mobile, Ad Hoc, and Wireless Networks (WiOpt 2017)*, pages 1–8, May 2017.

40 M. J. Farooq and Q. Zhu. Adaptive and resilient revenue maximizing dynamic resource allocation and pricing for cloud-enabled IoT systems. In *American Control Conference (ACC 2018)*, Milwaukee, WI, USA, Jun. 2018.

41 M. J. Farooq and Q. Zhu. On the secure and reconfigurable multi-layer network design for critical information dissemination in the Internet of battlefield things (IoBT). *IEEE Transactions on Wireless Communications*, 17 (4): 2618–2632, Apr. 2018.

42 M. J. Farooq, H. ElSawy, and M. S. Alouini. Modeling inter-vehicle communication in multi-lane highways: a stochastic geometry approach. In *IEEE 82nd Vehicular Technology Conference (VTC-Fall 2015)*, Boston, MA, USA, Sept. 2015.

43 M. J. Farooq, H. ElSawy, and M. S. Alouini. A stochastic geometry model for multi-hop highway vehicular communication. *IEEE Transactions on Wireless Communications*, 15 (3): 2276–2291, Mar. 2016.

44 M. J. Farooq, H. ElSawy, Q. Zhu, and M.-S. Alouini. Optimizing mission critical data dissemination in massive IoT networks. In *Workshop on Spatial Stochastic Models for Wireless Networks (SpaSWin, WiOpt 2017)*, Paris, France, May 2017.

45 M. Feily, A. Shahrestani, and S. Ramadass. A survey of botnet and botnet detection. In *3rd International Conference on Emerging Security Information, Systems, and Technologies*, pages 268–273, Jun. 2009.

46 G. Fettweis et al. The Tactile Internet: ITU-T Technology Watch Report, Aug. 2014.

47 Fog computing and the Internet of things: extend the cloud to where the things are. Cisco Inc. White Paper, 2015.

48 M. Frustaci, P. Pace, G. Aloi, and G. Fortino. Evaluating critical security issues of the IoT world: present and future challenges. *IEEE Internet of Things Journal*, 5 (4): 2483–2495, Aug. 2018.

49 P. Garcia Lopez, A. Montresor, D. Epema, A. Datta, T. Higashino, A. Iamnitchi, M. Barcellos, P. Felber, and E. Riviere. Edge-centric computing: vision and challenges. *SIGCOMM Computer Communication Review*, 45 (5): 37–42, Sept. 2015. ISSN 0146-4833.

50 I. M. Gelfand and S. V. Fomin. *Calculus of Variations (Dover Books on Mathematics)*. Dover Publications, 2000. ISBN 0486414485.

51 A. Gershkov and B. Moldovanu. Dynamic revenue maximization with heterogeneous objects: a mechanism design approach. *American Economic Journal: Microeconomics*, 1 (2): 168–198, 2009.

52 A. Gershkov and B. Moldovanu. *Dynamic Allocation and Pricing*. MIT Press, 2014.

53 H. Ghazzai, M. J. Farooq, A. Alsharoa, E. Yaacoub, A. Kadri, and M. Alouini. Green networking in cellular HetNets: a unified radio resource management framework with base station ON/OFF switching. *IEEE Transactions on Vehicular Technology*, 66 (7): 5879–5893, Jul. 2017.

54 R. Guedes, V. Furtado, and T. Pequeno. Multi-objective evolutionary algorithms and multiagent models for optimizing police dispatch. In *Proceedings of the IEEE International Conference on Intelligence and Security Informatics (ISI 2015)*, pages 37–42, May 2015.

55 B. Guo, Y. Liu, L. Wang, V. O. K. Li, J. C. K. Lam, and Z. Yu. Task allocation in spatial crowdsourcing: current state and future directions. *IEEE Internet of Things Journal*, 5 (3): 1749–1764, Jun. 2018.

56 X. Guo, R. Singh, T. Zhao, and Z. Niu. An index based task assignment policy for achieving optimal power-delay tradeoff in edge cloud systems. In *IEEE International Conference on Communications (ICC 2016)*, May 2016.

57 M. Haenggi. Outage, local throughput, and capacity of random wireless networks. *IEEE Transactions on Wireless Communications*, 8 (8): 4350–4359, Aug. 2009. ISSN 1536-1276. doi: 10.1109/TWC.2009.090105.

58 M. Haenggi, J. G. Andrews, F. Baccelli, O. Dousse, and M. Franceschetti. Stochastic geometry and random graphs for the analysis and design of wireless networks. *IEEE Journal on Selected Areas in Communications*, 27 (7): 1029–1046, Sept. 2009.

59 Y. Han, Y. Zhu, and J. Yu. Utility-maximizing data collection in crowd sensing: an optimal scheduling approach. In *Proceedings of the 12th Annual IEEE International Conference on Sensing, Communication, and Networking (SECON 2015)*, pages 345–353, Jun. 2015.

60 G. H. Hardy, J. E. Littlewood, and G. Polya. *Inequalities*. Cambridge University Press, Cambridge, 1934.

61 D. He, S. Chan, and M. Guizani. Drone-assisted public safety networks: the security aspect. *IEEE Communications Magazine*: 55 (8): 218–223, Aug 2017.

62 S. G. Henderson and A. J. Mason. *Ambulance Service Planning: Simulation and Data Visualisation*, pages 77–102. Springer US, Boston, MA, 2004.

63 J. K. Hunter and B. Nachtergaele. *Applied Analysis*, chapter The Contraction Mapping Theorem. World Scientific, 2001.

64 N. V. Juliadotter and K. K. R. Choo. Cloud attack and risk assessment taxonomy. *IEEE Cloud Computing*, 2 (1): 14–20, Jan. 2015.

65 E. Kalantari, M. Z. Shakir, H. Yanikomeroglu, and A. Yongacoglu. Backhaul-aware robust 3D drone placement in 5G+ wireless networks. In *IEEE International Conference on Communications (ICC 2017)*, Paris, France, May 2017.

66 M. Kaynia and N. Jindal. Performance of ALOHA and CSMA in spatially distributed wireless networks. In *2008 IEEE International Conference on Communication*, pages 1108–1112, May 2008.

67 S. F. Kazerooni and R. Rojas-Cessa. SRA: Slot reservation announcement scheme for medium access control of IEEE 802.11 crowded networks in emergency scenarios. In *IEEE International Conference on Communication (ICC 2017)*, Paris, France, May 2017.

68 E. Khorov, A. Kiryanov, and A. Lyakhov. Analysis of multiplexed streaming via periodic reservations of wireless channel. In *IEEE International Black Sea Conference on Communications and Networking (BlackSeaCom 2016)*, pages 1–5, Varna, Bulgaria, Jun. 2016.

69 T. Kim, D. M. Kim, N. Pratas, P. Popovski, and D. K. Sung. An enhanced access reservation protocol with a partial preamble transmission mechanism in NB-IoT systems. *IEEE Communications Letters*, 21 (10): 2270–2273, Oct. 2017.

70 J. F. C. Kingman. Markov population processes. *Journal of Applied Probability*, 6 (1): 1–18, 1969.

71 R. Knutson. Cell networks suffer outages in Harvey's wake. Technical report, Wall Street Journal, Aug. 2017. https://www.wsj.com/articles/cell-networks-suffer-outages-in-harveys-wake-1503792185.

72 M. Knysz, X. Hu, Y. Zeng, and K. G. Shin. Open WiFi networks: lethal weapons for botnets? In *Proceedings of IEEE International Conference on Computer Communication (INFOCOM 2012)*, pages 2631–2635, Orlando, FL, USA, Mar. 2012.

73 C. Kolias, G. Kambourakis, A. Stavrou, and J. Voas. DDoS in the IoT: Mirai and other botnets. *Computer*, 50 (7): 80–84, 2017.

74 H. W. Kuhn and B. Yaw. The Hungarian method for the assignment problem. *Naval Research Logistics Quarterly*: 83–97, 1955.

75 T. Lassen. Long-range RF communication: why narrowband is the de facto standard. Technical report, Texas Instruments, 2014.

76 Latency: The Achilles heel of cloud computing. Internap White Paper 877.THE.PNAP (877.843.7627), 2010.

77 Z. Li, S. Zozor, J. M. Drossier, N. Varsier, and Q. Lampin. 2D time-frequency interference modelling using stochastic geometry for performance evaluation in low-power wide-area networks. In *2017 IEEE International Conference on Communications (ICC)*, Paris, France, May 2017.

78 J. Lin, W. Yu, N. Zhang, X. Yang, H. Zhang, and W. Zhao. A survey on Internet of things: architecture, enabling technologies, security

and privacy, and applications. *IEEE Internet of Things Journal*, 4 (5): 1125–1142, Oct. 2017. ISSN 2327-4662.

79 Q. Liu, T. Han, and N. Ansari. Joint radio and computation resource management for low latency mobile edge computing. In *2018 IEEE Global Communications Conference (GLOBECOM)*, pages 1–7, Dec. 2018.

80 S. Lloyd. Least squares quantization in PCM. *IEEE Transactions on Information Theory*, 28 (2): 129–137, Mar. 1982.

81 E. F. Long, E. Nohdurft, and S. Spinler. Spatial resource allocation for emerging epidemics: a comparison of greedy, myopic, and dynamic policies. *Manufacturing & Service Operations Management*, 20 (2): 181–198, May 2018. ISSN 1526-5498.

82 Y. Lu, S. T. Maguluri, M. S. Squillante, and C. W. Wu. Risk-based dynamic allocation of computing resources. *SIGMETRICS Performance Evaluation Review*, 44 (2): 27–29, Sept. 2016. ISSN 0163-5999.

83 H. Luo, Z. Chen, J. Li, and A. V. Vasilakos. Preventing distributed denial-of-service flooding attacks with dynamic path identifiers. *IEEE Transactions on Information Forensics and Security*, 12 (8): 1801–1815, Aug. 2017.

84 Y. Luo, L. Gao, and J. Huang. Spectrum reservation contract design in TV white space networks. *IEEE Transactions on Cognitive Communications and Networking*, 1 (2): 147–160, Jun. 2015.

85 J. Luo, L. Rao, and X. Liu. Data center energy cost minimization: a spatio-temporal scheduling approach. In *Proceedings of the IEEE Conference on Computer Communications (INFOCOM 2013)*, pages 340–344, Apr. 2013.

86 M. Lus, R. Oliveira, R. Dinis, and L. Bernardo. A novel reservation-based MAC scheme for distributed cognitive radio networks. *IEEE Transactions on Vehicular Technology*, 66 (5): 4327–4340, May 2017.

87 J. Lyu, Y. Zeng, R. Zhang, and T. J. Lim. Placement optimization of uav-mounted mobile base stations. *IEEE Communications Letters*, 21 (3): 604–607, Mar. 2017.

88 M. Maciejewski, J. Bischoff, and K. Nagel. An assignment-based approach to efficient real-time city-scale taxi dispatching. *IEEE Intelligent Systems*, 31 (1): 68–77, Jan. 2016.

89 A. Malik, R. Maciejewski, S. Towers, S. McCullough, and D. S. Ebert. Proactive spatiotemporal resource allocation and predictive visual analytics for community policing and law enforcement. *IEEE Transactions on Visualization and Computer Graphics*, 20 (12): 1863–1872, Dec. 2014.

90 J. Marro and R. Dickman. *Nonequilibrium Phase Transitions in Lattice Models*, chapter The contact process. Cambridge University Press, New York, 1999.

91 S. Marrone and R. Nardone. Automatic resource allocation for high availability cloud services. *Procedia Computer Science*, 52: 980–987, 2015. ISSN 1877-0509.

92 A. Marttinen, A. M. Wyglinski, and R. Jntti. Statistics-based jamming detection algorithm for jamming attacks against tactical MANETs. In *IEEE Military Communications Conference (MILCOM 2014)*, Baltimore, MD USA, Oct 2014.

93 H. Menouar, I. Guvenc, K. Akkaya, A. S. Uluagac, A. Kadri, and A. Tuncer. UAV-enabled intelligent transportation systems for the smart city: applications and challenges. *IEEE Communications Magazine*, 55 (3): 22–28, Mar. 2017.

94 F. Miao, S. Han, S. Lin, Q. Wang, J. A. Stankovic, A. Hendawi, D. Zhang, T. He, and G. J. Pappas. Data-driven robust taxi dispatch under demand uncertainties. *IEEE Transactions on Control Systems Technology*, 27 (1): 175–191, Jan. 2019.

95 Z. Miao, Y. H. Liu, Y. Wang, and R. Fierro. Formation control of quadrotor UAVs without linear velocity measurements. In *Proceedings of the 18th International Conference on Advanced Robotics (ICAR 2017)*, pages 179–184, Hong Kong, China, Jul. 2017.

96 Z. Miao, Y. Wang, and R. Fierro. Collision-free consensus in multi-agent networks: a monotone systems perspective. *Automatica*, 64 (Supplement C): 217–225, 2016.

97 P. Moriuchi and S. Chohan. Mirai-variant IoT botnet used to target financial sector in January 2018. Insikt Group, Apr. 2018.

98 M. Mozaffari, W. Saad, M. Bennis, and M. Debbah. Efficient deployment of multiple unmanned aerial vehicles for optimal wireless coverage. *IEEE Communications Letters*, 20 (8): 1647–1650, Aug. 2016.

99 M. Mozaffari, W. Saad, M. Bennis, and M. Debbah. Optimal transport theory for power-efficient deployment of unmanned aerial vehicles. In *IEEE International Conference on Communications (ICC 2016)*, Kuala Lumpur, Malaysia, May 2016.

100 M. Mozaffari, W. Saad, M. Bennis, and M. Debbah. Mobile unmanned aerial vehicles (UAVs) for energy-efficient internet of things communications. *IEEE Transactions on Wireless Communications*, 16 (11): 7574–7589, Nov. 2017.

101 R. B. Myerson. Optimal auction design. *Mathematics of Operations Research*, 6 (1): 58–73, Feb. 1981. ISSN 0364-765X. doi: 10.1287/moor.6.1.58.

102 R. B. Myerson. *Mechanism Design*. Palgrave Macmillan UK, London, 1989.

103 K. E. Nolan, W. Guibene, and M. Y. Kelly. An evaluation of low power wide area network technologies for the internet of things. In *Proceedings of International Wireless Communications and Mobile Computing Conference (IWCMC 2016)*, pages 439–444, Cyprus, Paphos, Sept. 2016.

104 NYC OpenData, NYC Wi-Fi Hotspot Locations, https://data
.cityofnewyork.us/Social-Services/NYC-Wi-Fi-Hotspot-Locations/a9we-
mtpn.

105 R. Olfati-Saber. Flocking for multi-agent dynamic systems: algorithms and
theory. *IEEE Transactions on Automatic Control*, 51 (3): 401–420, Mar.
2006.

106 T. Omitola and G. Wills. Towards mapping the security challenges of the
Internet of things (IoT) supply chain. *Procedia Computer Science*, 126:
441–450, 2018.

107 M. R. Palattella, M. Dohler, A. Grieco, G. Rizzo, J. Torsner, T. Engel, and
L. Ladid. Internet of things in the 5G era: enablers, architecture, and business models. *IEEE Journal on Selected Areas in Communications*, 34 (3):
510–527, Mar. 2016.

108 D. P. Palomar and M. Chiang. A tutorial on decomposition methods
for network utility maximization. *IEEE Journal on Selected Areas in
Communications*, 24 (8): 1439–1451, Aug. 2006. ISSN 0733-8716. doi:
10.1109/JSAC.2006.879350.

109 R. Pastor-Satorras, C. Castellano, P. Van Mieghem, and A. Vespignani.
Epidemic processes in complex networks. *Reviews of Modern Physics*, 87:
925–979, Aug. 2015. doi: 10.1103/RevModPhys.87.925.

110 R. Pastor-Satorras and A. Vespignani. Epidemic spreading in scale-free
networks. *Physical Review Letters*, 86: 3200–3203, Apr. 2001. http://link
.aps.org/doi/10.1103/PhysRevLett.86.3200.

111 K. Pelechrinis, M. Iliofotou, and S. V. Krishnamurthy. Denial of service
attacks in wireless networks: the case of jammers. *IEEE Communication
Surveys and Tutorials*, 13 (2): 245–257, Second Quarter 2011.

112 T. Reed, J. Geis, and S. Dietrich. SkyNET: a 3G-enabled mobile attack
drone and stealth botmaster. In *Proceedings of the 5th USENIX Conference
on Offensive Technologies*, WOOT'11, Berkeley, CA, USA, 2011. USENIX
Association.

113 P. Reverdy and D. E. Koditschek. Mobile robots as remote sensors for
spatial point process models. In *IEEE/RSJ International Conference on
Intelligent Robots and Systems (IROS 2016)*, pages 2847–2852, Daejeon,
Korea, Oct. 2016.

114 S. M. Ross. *Stochastic Processes*. Wiley, 1996.

115 R. O. Saber and R. M. Murray. Flocking with obstacle avoidance: coop-
eration with limited communication in mobile networks. In *42nd IEEE
International Conference on Decision and Control (CDC 2003)*, Volume 2,
pages 2022–2028, Dec. 2003.

116 D. Sahinel, C. Akpolat, F. Sivrikaya, and S. Albayrak. An agent-based network resource management concept for smart city services. In *Proceedins
of the 14th Annual Conference on Wireless On-demand Network Systems
and Services (WONS 2018)*, pages 129–132, Feb. 2018.

117 R. Sanchez, J. Evans, and G. Minden. Networking on the battlefield: challenges in highly dynamic multi-hop wireless networks. In *IEEE Military Communications Conference (MILCOM 1999)*, 1999.

118 J. Santos, T. Vanhove, M. Sebrechts, T. Dupont, W. Kerckhove, B. Braem, G. V. Seghbroeck, T. Wauters, P. Leroux, S. Latre, B. Volckaert, and F. D. Turck. City of things: enabling resource provisioning in smart cities. *IEEE Communications Magazine*, 56 (7): 177–183, Jul. 2018.

119 J. Santos, T. Wauters, B. Volckaert, and F. De Turck. Resource provisioning for IoT application services in smart cities. In *Proceedings of the 13th International Conference on Network and Service Management (CNSM 2017)*, pages 1–9, Nov. 2017.

120 J. Sanz, C.-Y. Xia, S. Meloni, and Y. Moreno. Dynamics of interacting diseases. *Physical Review X*, 4: 041005, Oct. 2014. http://link.aps.org/doi/10.1103/PhysRevX.4.041005.

121 S. Sarkar, S. Chatterjee, and S. Misra. Assessment of the suitability of fog computing in the context of Internet of things. *IEEE Transactions on Cloud Computing*, 6 (1): 46–59, Jan. 2018.

122 M. Schwager, M. P. Vitus, S. Powers, D. Rus, and C. J. Tomlin. Robust adaptive coverage control for robotic sensor networks. *IEEE Transactions on Control of Network Systems*, 4 (3): 462–476, Sept. 2017.

123 K. T. Seow, N. H. Dang, and D. Lee. A collaborative multiagent taxi-dispatch system. *IEEE Transactions on Automation Science and Engineering*, 7 (3): 607–616, Jul. 2010.

124 S. A. W. Shah, T. Khattab, M. Z. Shakir, and M. O. Hasna. A distributed approach for networked flying platform association with small cells in 5G+ networks. In *IEEE Global Communications Conference (Globecom 2017)*, Singapore, Dec. 2017.

125 S. A. W. Shah, T. Khattab, M. Z. Shakir, and M. O. Hasna. Association of networked flying platforms with small cells for network centric 5G+ C-RAN. In *Proceedings of IEEE International Symposium on Personal, Indoor and Mobile Radio Communications (PIMRC 2017)*, Montreal, Canada, Sept. 2017.

126 H. Shah-Mansouri and V. W. S. Wong. Hierarchical fog-cloud computing for IoT systems: a computation offloading game. *IEEE Internet of Things Journal*, 5 (4): 3246–3257, Aug. 2018.

127 H. Shen and L. Chen. Resource demand misalignment: an important factor to consider for reducing resource over-provisioning in cloud datacenters. *IEEE/ACM Transactions on Networking*, 26 (3): 1207–1221, Jun. 2018.

128 W. Shi and S. Dustdar. The promise of edge computing. *Computer*, 49 (5): 78–81, May 2016.

129 W. Shi, J. Cao, Q. Zhang, Y. Li, and L. Xu. Edge computing: vision and challenges. *IEEE Internet of Things Journal*, 3 (5): 637–646, Oct. 2016.

130 W. Song, Z. Xiao, and Q. Chen. Dynamic resource allocation using virtual machines for cloud computing environment. *IEEE Transactions on Parallel and Distributed Systems*, 24: 1107–1117, 2013. ISSN 1045-9219.

131 V. B. C. Souza, W. Ramrez, X. Masip-Bruin, E. Marn-Tordera, G. Ren, and G. Tashakor. Handling service allocation in combined fog-cloud scenarios. In *IEEE International Conference on Communications (ICC 2016)*, pages 1–5, May 2016.

132 J. A. Stankovic. Research directions for the internet of things. *IEEE Internet of Things Journal*, 1 (1): 3–9, Feb. 2014.

133 D. Stoyan, W. S. Kendall, and J. Mecke. *Stochastic Geometry and Its Applications. Wiley Series in Probability and Mathematical Statisitics.* Wiley, Chichester, W. Sussex, New York, 1987. ISBN 0-471-90519-4.

134 H. Tan, Z. Han, X. Li, and F. C. M. Lau. Online job dispatching and scheduling in edge-clouds. In *IEEE Conference on Computer Communications (INFOCOM 2017)*, May 2017.

135 C. S. Tang. Perspectives in supply chain risk management. *International Journal of Production Economics*, 103 (2): 451–488, 2006. ISSN 0925-5273.

136 M. Tanha, D. Sajjadi, F. Tong, and J. Pan. Disaster management and response for modern cellular networks using flow-based multi-hop device-to-device communications. In *IEEE 84th Vehicular Technology Conference (VTC-Fall 2016)*, Montral, Canada, Sept. 2016.

137 A. Tannenbaum. Why do IoT companies keep building devices with huge security flaws? Harvard Business Review, Apr. 2017.

138 L. Tong, Y. Li, and W. Gao. A hierarchical edge cloud architecture for mobile computing. In *35th Annual IEEE International Conference on Computer Communications (INFOCOM 2016)*, pages 1–9, Apr. 2016.

139 H.-R. (Debbie) Tsai, Y. Shoukry, M. K. Lee, and V. Raman. Towards a socially responsible smart city: dynamic resource allocation for smarter community service. In *Proceedings of the 4th ACM International Conference on Systems for Energy-Efficient Built Environments (BuildSys 2017)*, pages 13:1–13:4, New York, NY, USA, 2017. ACM. ISBN 978-1-4503-5544-5. doi: 10.1145/3137133.3137163.

140 L. Vandenberghe, S. P. Boyd, and S. Boyd. *Convex Optimization.* Cambridge University Press, Cambridge, UK, 2004.

141 P. Van Mieghem. The *N*-intertwined SIS epidemic network model. *Computing*, 93 (2): 147–169, 2011.

142 C. Villani. *Optimal Transport: Old and New.* Grundlehren der mathematischen Wissenschaften. Springer, Sept. 2008.

143 G. Vormayr, T. Zseby, and J. Fabini. Botnet communication patterns. *IEEE Communication Surveys and Tutorials*, 19 (4): 2768–2796, Fourth Quarter 2017.

144 J. Wang, H. Zhong, Q. Xia, and C. Kang. Optimal planning strategy for distributed energy resources considering structural transmission

cost allocation. *IEEE Transactions on Smart Grid*, 9 (5): 5236–5248, Sept. 2018.

145 S. Weber, J. G. Andrews, and N. Jindal. An overview of the transmission capacity of wireless networks. *IEEE Transactions on Communications*, 58 (12): 3593–3604, Dec. 2010. ISSN 0090-6778. doi: 10.1109/TCOMM.2010.093010.090478.

146 L. Xiao, M. Johansson, and S. P. Boyd. Simultaneous routing and resource allocation via dual decomposition. *IEEE Transactions on Communications*, 52 (7): 1136–1144, July 2004. ISSN 0090-6778. doi: 10.1109/TCOMM. 2004.831346.

147 H. Xu and B. Li. Dynamic cloud pricing for revenue maximization. *IEEE Transactions on Cloud Computing*, 1 (2): 158–171, July 2013.

148 J. Xu, R. Rahmatizadeh, L. Bölöni, and D. Turgut. Taxi dispatch planning via demand and destination modeling. In *Proceedings of IEEE 43rd Conference on Local Computer Networks (LCN 2018)*, pages 377–384, Oct. 2018.

149 Q. Yang, D. An, R. Min, W. Yu, X. Yang, and W. Zhao. On optimal PMU placement-based defense against data integrity attacks in smart grid. *IEEE Transactions on Information Forensics and Security*, 12 (7): 1735–1750, Jul. 2017.

150 Y. Yang, L. Wu, G. Yin, L. Li, and H. Zhao. A survey on security and privacy issues in Internet-of-Things. *IEEE Internet of Things Journal*, 4 (5): 1250–1258, Oct. 2017.

151 J. Yao and N. Ansari. Reliability-aware fog resource provisioning for deadline-driven IoT services. In *2018 IEEE Global Communications Conference (GLOBECOM)*, pages 1–6, Dec. 2018.

152 J. Yao and N. Ansari. QoS-aware fog resource provisioning and mobile device power control in iot networks. *IEEE Transactions on Network and Service Management*, 16 (1): 167–175, Mar. 2019.

153 A. Yousefpour, G. Ishigaki, and J. P. Jue. Fog computing: towards minimizing delay in the internet of things. In *IEEE International Conference on Edge Computing (EDGE 2017)*, Jun. 2017.

154 A. Yousefpour, G. Ishigaki, R. Gour, and J. P. Jue. On reducing IoT service delay via fog offloading. *IEEE Internet of Things Journal*, 5 (2): 998–1010, Apr. 2018.

155 W. Yu and R. Lui. Dual methods for nonconvex spectrum optimization of multicarrier systems. *IEEE Transactions on Communications*, 54 (7): 1310–1322, Jul. 2006. ISSN 0090-6778. doi: 10.1109/TCOMM.2006.877962.

156 A. Zanella, N. Bui, A. Castellani, L. Vangelista, and M. Zorzi. Internet of things for smart cities. *IEEE Internet of Things Journal*, 1 (1): 22–32, Feb. 2014.

157 Y. Zeng, R. Zhang, and T. J. Lim. Wireless communications with unmanned aerial vehicles: opportunities and challenges. *IEEE Communications Magazine*, 54 (5): 36–42, May 2016.

158 Q. Zhang and F. H. P. Fitzek. *Mission Critical IoT Communication in 5G*, pages 35–41. Springer International Publishing, Cham, 2015.

159 H. Zhang, B. Li, H. Jiang, F. Liu, A. V. Vasilakos, and J. Liu. A framework for truthful online auctions in cloud computing with heterogeneous user demands. In *2013 Proceedings IEEE INFOCOM*, pages 1510–1518, Apr. 2013.

160 Q. Zhang, Q. Zhu, and R. Boutaba. Dynamic resource allocation for spot markets in cloud computing environments. In *4th IEEE International Conference on Utility and Cloud Computing*, pages 178–185, Dec. 2011.

161 Q. Zhu, M. Y. S. Uddin, Z. Qin, and N. Venkatasubramanian. Upload planning for mobile data collection in smart community internet-of-things deployments. In *Proceedings of IEEE International Conference on Smart Computing (SMARTCOMP 2016)*, pages 1–8, May 2016.

162 Q. Zhu, M. Y. S. Uddin, N. Venkatasubramanian, and C. Hsu. Spatiotemporal scheduling for crowd augmented urban sensing. In *Proceedings of IEEE Conference on Computer Communications (INFOCOM 2018)*, Apr. 2018.

163 J. R. Zipkin, M. B. Short, and A. L. Bertozzi. Cops on the dots in a mathematical model of urban crime and police response. *Discrete and Continuous Dynamical Systems - B*, 19 (1531–3492): 1479, Jul. 2014.

Index

Resource Management for On-Demand Mission-Critical Internet of Things Applications, First Edition.
Junaid Farooq and Quanyan Zhu.
© 2021 John Wiley & Sons, Inc. Published 2021 by John Wiley & Sons, Inc.

Printed and bound by CPI Group (UK) Ltd, Croydon, CR0 4YY